D0437751

ATTENTION ALL PASSENGERS

ATTENTION ALL PASSENGERS

THE AIRLINES' DANGEROUS DESCENT—AND HOW TO RECLAIM OUR SKIES

WILLIAM J. McGEE

HARPER
An Imprint of HarperCollins*Publishers*
www.harpercollins.com

FIRST EDITION

Designed by William Ruoto

Library of Congress Cataloging-in-Publication Data has been applied for.

ISBN: 978-0-06-208837-6

12 13 14 15 16 OV/RRD 10 9 8 7 6 5 4 3 2 1

FOR NICK

Contents

Author's Note

THE AUTHOR DOES NOT RECEIVE any income from airlines or other travel suppliers and does not invest or own shares in any travel companies. He also does not collect or redeem airline frequent flyer mileage or travel loyalty points. Although he worked for three U.S. airlines between 1985 and 1992, all three were financially liquidated and he receives no pension or benefits. Furthermore, he did not accept any compensatory flights or other free services, accommodations, or gifts from airlines or other travel suppliers while researching this book. All quotes in the book that are not cited with an endnote originated in a series of interviews conducted by the author between January and August 2011.

Acknowledgments

FOREMOST, I'D LIKE TO EXPRESS gratitude to my family, who made this and so much else possible: my parents; my brothers and sisters—Bob, Judy, Kathy, Rich, Mike, Janice, Pat, Margie, John, and Nancy; and their families. And Nick, for giving up the dining room table.

At HarperCollins, I've been very lucky to work with Colleen Lawrie, who always exhibits a fine eye for editing, as well as endless patience and good humor. Special thanks to Shannon Ceci, Beth Harper, Katie O'Callaghan, and Mark Ferguson, in addition to Matt Inman and Julia Cheiffetz for early assistance.

I have also been lucky to undertake this journey with Rob Weisbach and David Groff, both of whom have never failed to provide encouragement, support, and wise counsel.

Unflagging support above and beyond the call were provided by Irene Zutell and Larry Bleidner, and Andrew "Andy Five Angels" Postman. Many thanks to Kevin Mitchell, John Conley, Gabe Bruno, John Goglia, and Bogdan Dzakovic. Invaluable assistance was provided by Tom Wojtaszek, Wes Gill, Marlys Harris, Judy Stone, Barbara S. Peterson, Nick Galifianakis, Mario Santamaria, and Beau Brendler.

Among my aviation friends and colleagues, there are way more than a few who cannot be named, as well as Lieutenant Kevin Coughlin, Jackie Sullivan, Ray Hodgert, and the late Ron

Bello. The many others know who they are. Any measure of realism is due to their efforts; any inaccuracies are solely mine. In addition, this book was aided immeasurably by the cooperation of the brave men and women who are FAA, TSA, and airline whistle-blowers—I'm humbled by the trust they placed in me.

Special thanks to Linda Burbank, Debra D'Agostino, Lizz Dinnigan, Pam Fehl, Sue Juliano, Doug Love and Lisa Pulitzer, Donna Marino Wilkins, Michele McDonald, Alex Postman, and Stacy Small. And Dawn Barclay, Mercedes Cardona, Phyllis Fine, Mike Milligan, and Melissa Ng. At Hofstra, I'm very grateful for the support from my colleagues Joe Fichtelberg, Julia Markus, Erik Brogger, Martha McPhee, Pat and Paul Navarra, Rich Pioreck, Linda Reesman, Phillis Levin, Craig Rustici, Denise LoMonaco, Paula Curci, Ginny Greenberg, Meghan Curley, and Connie Roberts. And, of course, my talented students, particularly those in Hofstra Writers (please note: an entire book that does not contain the words *amazing* or *awesome*). From the Columbia MFA program, I'm especially indebted to Lis Harris for her unswerving faith and commitment, as well as Stephen Koch for early and much-appreciated support. And thanks to Chuck Bell, Ellen Bloom, and Lisa Freeman at Consumers Union and Ben Abramson at USAToday.com.

Finally, I'm grateful to Hofstra College of Liberal Arts and Sciences for a much-needed Faculty Research and Development Grant that assisted with the research and travel necessary for this book.

ATTENTION ALL PASSENGERS

Prologue: Flying Sucks

I USED TO LOVE THE airlines. I loved everything about flying. In fact, aviation seems to run in my blood. I have relatives who have piloted airplanes, fixed airplanes, and ridden airplanes into combat, and I have quite a few siblings and cousins who have toiled for airlines over the years. Way back in June 1927 my grandfather Bill McMullen snuck my mother out of school and escorted her onto a trolley car so they could travel from Queens into Manhattan. She was six years old and had spent most of the winter in a sanitarium fighting for her life against diphtheria, so my guess is that he left his other kids behind so he could remind her just how sweet life can be. He hoisted his little girl onto his shoulders as she watched Colonel Charles Lindbergh, the world's most famous pilot, pass by in a ticker-tape parade. They were celebrating the Lone Eagle's triumphant return from Paris, a city my grandfather—and namesake—had visited just a few years earlier during World War I with the American Expeditionary Forces.

It's a year worth considering. The American Century was unfolding the way Broadway unfolded before Lucky Lindy during that parade. The possibilities must have seemed limitless: America was moving faster, further, higher. Despite the ominous stock bubble and the nascent march of fascism, the country clearly had faith in its workforce, in its technology, and in itself. It's no coincidence that 1927 was also when a blue-blood Yalie and former

U.S. Navy pilot named Juan Trippe founded what would become the world's all-time greatest airline, Pan Am.

More than six decades later, in 1991, I was the operations system control manager on duty when the final Pan Am Shuttle flight rolled up to the gate at LaGuardia Airport in New York. In 1985 I had taken what I thought would be a summer job in the airline industry, which quickly led to working for four different carriers, and in many ways, it's as if I never left. After twenty-seven years, I'm still immersed in the business, though now I write and advocate about it rather than work in it. As the old-timers say, I've got Jet A-1 fuel coursing through my veins.

I really did love the business. I loved airplanes. I worked in the airlines, went flying on my days off, served in the Air Force Auxiliary, vacationed at air shows, and even stayed up nights writing about AirFair, my own fictional air carrier. As Ed Acker, the former CEO of Air Florida and Pan Am, aptly noted, "Once you get hooked on the airline business, it's worse than dope." Amen.

And so, like millions of other Americans, it pains me to see what's happened to what was at one time the exhilarating experience of boarding a flight. Today, commercial flying sucks. And everyone knows it.

The first time I traveled to an airport I was five years old. Most of my very large family piled into our 1965 Mercury Colony Park station wagon and my brother and I had to be extra careful crawling over the tailgate to settle into the rear-facing third seat. That was because we were wearing dark suits, white shirts, and ties. My sisters were wearing dresses and shiny shoes. We were en route to John F. Kennedy International Airport. *And we weren't even flying!* It was 1967 and my oldest brother was home from Fort Gordon, in Georgia, for Thanksgiving leave. Think about that. We were just visiting an airport, one of dozens of families waiting in the arrivals hall as a planeload of shaved GIs filed through with duffel bags slung over their shoulders. And we were dressed as if it were Easter Sunday.

I consider that every time I'm crammed in next to some guy who crosses his bare legs against my tray table as I count the curly little hairs on the knuckle of his big toe and munch on the Blueberry Pomegranate Trail Mix Crunch that is just four dollars on US Airways. In researching this book, I spoke to a lot of veteran travelers who whined about the good old days romanticized on TV in *Mad Men* and *Pan Am*—you know, when flyers didn't dress in shorts and tank tops and flip-flops, employ stage voices to speak on cell phones in crowded jet bridges, hit old ladies in the noggin cramming overstuffed carry-ons into overhead bins, or engage in fisticuffs with flight attendants. In other words, they were civilized.

Let's be clear. We asked for democracy, for the eradication of that dated term *jet set*, for air travel to become accessible to the masses, for flying to become a God-given constitutional right like bearing arms or blogging in all uppercase letters with no punctuation. We asked for what the Universal Declaration of Human Rights calls "freedom of movement." And the 1978 Airline Deregulation Act gave us just that. It took the government out of the airline scheduling business and empowered the free market to determine where and how often carriers would fly, and how much they would charge passengers. So for some the perception is that behavior that used to be confined to other locations has become commonplace at thirty-five thousand feet.

But it doesn't have to be this way. The masses are also allowed into shopping malls and restaurants, theaters and arenas, schools and offices, parks and libraries and houses of worship. Somehow those venues don't erupt into Argentinean *futbol* riots. One would think the terms *subway rage* or *bus rage* would have entered the lexicon before *air rage* did. Today, riding Amtrak offers a better customer experience than flying on an airline. Heck, riding Greyhound offers a better customer experience. No, flying sucks because the airlines have rapidly deteriorated, due to a handful of greedy executives who have sucked the civility out of our own publicly funded airports and airways, and lax government regulators who have allowed customer service, security, and even safety to decline—all in the name of worshipping the free market, which of course is anything but free for customers. Carriers refuse to put money into improving their operations, and we're partly to blame for allowing it all to unravel in such a disheartening way.

Air travel has become a commodity, and the airlines themselves an oligopoly that has carved up hub airports the way cable television providers have carved up zip codes. Consequently airline execs do not care about customers in the way a service industry should. Contacting an airline has become akin to calling your cable company when there's an outage—good luck with that. In fact, the chief executive of the Irish low-cost airline Ryanair was quoted a few years back as saying: "Air transport is just a glorified bus operation." These days, many passengers might not have a problem with that view. The problem is that it's a *poorly run* bus operation.

Over nearly three decades, I've been given a unique micro and macro view of the airline experience. I began my career by loading bags out on the tarmac and then moved into ground operations and flight operations management for third-tier players—Overseas National Airways, Ogden Allied, Tower Air. Then I worked for Pan Am. Eventually, I left aviation and began writing about the big picture—the CEOs, bankruptcies, mergers, start-ups, shutdowns, strikes, lawsuits, accidents, and the terrorism. And in recent years as the aviation consultant to nonprofit Consumers Union, I've become a passenger advocate, testifying before Congress and speaking out on customer service and safety. In May 2010, Secretary of Transportation Raymond

LaHood selected me as the consumer advocate on the Future of Aviation Advisory Committee, and I spent seven months debating these customer service and safety issues with airline executives and government regulators.

When my service on the FAAC ended, I decided to take a fresh look and embark on a journey of discovery about an industry I thought I knew. This book chronicles that journey, which took me from reservations centers in India to the lost luggage center in Alabama, from outsourced maintenance hangars to DOT headquarters, from the Boeing assembly line in Seattle to the aircraft boneyard in the California desert. My goal was to probe and inquire and investigate, all in the name of explaining this tremendously oversized and complex spectacle to average Americans. Most passengers have no idea how much has changed so quickly, and how little is transparent about the airline industry.

The term *airline* means many things to many people, but all can agree an airline traditionally contains three components: it owns airplanes, fixes airplanes, and flies airplanes. Today, most major carriers are distancing themselves from all three pursuits. Banks and leasing firms own most of the aircraft. And a radical sea change has occurred in airline maintenance. Nearly all U.S. carriers—with the lone exception of American Airlines—have elected to farm out their maintenance and repairs, both inside and outside the United States, thereby preventing the Federal Aviation Administration from providing adequate oversight. And today many airlines are even outsourcing flying itself, by engaging in marketing deals that allow carriers to label other airlines' flights as their own, especially on regional airplanes.

The first airline ticket was sold in 1914, and deregulation took effect nearly thirty-four years ago, but many of the most startling changes in the industry's history have quietly occurred within the last decade. Consider these facts:

- U.S. airlines generated $21.5 billion in 2010 from "ancillary" fees for services such as checking baggage.
- On any given day, there are six billion airfares loaded into computer reservations systems, fueling arcane and often nonsensical pricing mechanisms.
- Now 53 percent of all airline flight departures are operated by the major carriers' regional partners, such as Comair, Colgan, and Chautauqua.
- More than half of all frequent flyer mileage is now earned without ever boarding an airplane.
- Since 1978, 189 domestic airlines have filed for bankruptcy.
- The heads of the ten largest airlines in the United States earned $38,907,562 in salary and bonuses in 2010.
- About 71 percent of all U.S. airline heavy maintenance is now outsourced, in many cases to developing nations.
- It costs $700 million annually for air marshals to protect only 2 percent of U.S. flights.
- The aviation industry is responsible for 2–5 percent of climate change, and the industry's CO_2 output will grow 3–4 percent annually.

My friend Kevin Mitchell, who lobbies for passenger rights through the Business Travel Coalition, says airlines are engaged "in a mad race to the bottom on costs"—and that race is playing itself out through service, security, and safety. Mitchell also says, "The airlines are a proxy for American decline." I'm thoroughly convinced of that as well.

Ostensibly this book is about airlines, but in writing it, I've come to see that it's about much more—that the issues it addresses speak to all of twenty-first-century America. The airline story has become America's story: while we're entertained with bread and circuses, good jobs are downsized, outsourced, and offshored, the disconnect grows between service companies and their customers, Corporate America purchases government influence wholesale, federal regulators refuse to properly oversee our safety and security, and the financial chasm widens between senior executive "haves" and average worker "have-nots."

Nicole Piasecki, vice president for Boeing and a fellow member of the FAAC, speaks passionately about globalism and her vision for a unified world economy. I respect her immensely and understand these arguments, and I strive to stay engaged in a world that is rapidly shrinking. But I remain an American. This is where I live, work, write, raise my son, vote, and pay taxes. I adhere to American laws, support an American infrastructure, and expect my elected officials to prioritize national interests above foreign interests. For me, the bell tolls much more strongly in Minneapolis and Miami than in Beijing and San Salvador. And when Airbus finally buys up Boeing and Seattle goes the way of Detroit, are we supposed to pretend it won't be a bad thing for America because a handful of investors made out okay? Or do we just look forward to the next McDonald's hiring day and the golden fifty thousand openings that need filling?

That mad race to the bottom is already having ill effects. Right now the U.S. airline industry has a stellar safety record—but it needs to be taken in context. There are troubling warning signs—incidents and accidents that indicate standards are slipping. As Mitchell notes, on the day before the Deepwater Horizon explosion in April 2010, there had been a 100 percent safety record on oil rigs in the Gulf of Mexico, but now we know how a lack of government oversight can facilitate tragedy. Similarly, the major financial companies responsible for the 2008 Wall Street meltdown were enjoying financial ratings of AA or better at the time of their collapses and/or government bailouts. Indeed, airline safety is poised to be the next Deepwater Horizon catastrophe, Enron fiasco, or Fannie Mae/Freddie Mac calamity. Only this time we're all aware of it, with plenty of advance warning.

Just as war produces collateral damage, so too does capitalism. Economists refer to airlines "exiting the market" and jobs "going away." They're employing Orwellian tactics to soften the blows, but the airline industry battlefield has been littered with real pain. We've had no meaningful national discussion about the bankruptcies and jobs and security and safety and cutting corners—in more than thirty years. The time has come. As President Franklin D. Roosevelt told the Democratic National Convention in 1936, "Better the occasional faults of a government that lives in a spirit of charity than the consistent omissions of a government frozen in the ice of its own indifference."

Yet when we suggest that U.S. airplanes be fixed in the United States or workers be given a living wage or passengers be treated with respect or carbon footprints be curtailed, we're condescendingly told we don't understand the complexities, the Global Economy, the way the world works. But we *do* understand. As George Bernard Shaw stated, "The reasonable man adapts himself to the world; the unreasonable one persists to adapt the world to himself. Therefore all progress depends on the unreasonable man."

Ronald Reagan was lauded for two things above all: downsizing government and simplifying complex issues. So for a moment let's follow his example. Here's an easy way to consider if you truly want to get government completely off corporate backs. Don't view this as an abstract political, economic, or academic concept worthy of a robust debate. Instead make it personal. Take whomever you love most—adult, child, pet—and then strap that loved one into a pressurized aluminum tube, to be hurtled through the troposphere at four-fifths the speed of sound, roughly 550 miles per hour. And all the while, trust that this airplane was serviced properly, with no shortcuts, in a suitable facility with competent, licensed mechanics who understood their work and were screened by security and for alcohol and drug use.

Trust that the crew was trained properly and is earning a living wage. Trust completely in the oversight of the airline executives du jour who show loyalty only to shareholders and have golden parachutes for themselves.

Then ask yourself if you wouldn't maybe like a few inspectors to double-check their efforts. You may find that you don't have that much faith in Corporate America after all.

1

Sit Down and Shut Up or We'll Turn This Plane Around: Why Airline Service Has Collapsed and Air Rage Is Soaring

> Hi, I'm Jenn, your virtual assistant for the Alaska Airlines Web site. If you need help or have a question, simply type it below.
>
> —*AlaskaAir.com*

FOR MANY AIRLINE PASSENGERS, A big fear is that our checked luggage will turn up in a place most of us have never heard of—a place like, say, Scottsboro, Alabama. In fact, that's exactly what has happened with millions of suitcases over the years. According to federal regulations, the airlines have ninety days to reunite passengers and lost bags, and they usually manage to do it. However, on Day 91 a large truck will tote the unwanted belongings off to the Unclaimed Baggage Center. Before long, your iPod, your paperback Harlequin, and even your underwear will be up for sale.

You have to really work to find the Unclaimed Baggage Center (UBC), because Scottsboro is not convenient for anyone outside of northeastern Alabama (and is primarily known as the site of the infamous and racially charged "Scottsboro Boys" trial in the 1930s). The nearest commercial airport is a distant forty miles away in Huntsville. The UBC has been dispatching trucks to airline bag facilities since 1970 and now works exclusively with all major domestic carriers. Once the goodies arrive sight unseen in Scottsboro, they're unlocked, unpacked, sorted, and cleaned (UBC boasts of laundering more loads than anyone in Alabama); then 100 employees stock 5,000 to 7,000 items on the shelves every day. Here's the breakdown: 40 percent of what's found inside the bags goes to charity, 30 percent is recycled or tossed out, and the remaining 30 percent fills the 40,000-square-foot retail store in Scottsboro.

Talk about a niche market. There are a few wannabes along Willow Street, but you can't miss UBC's giant neon suitcase. Since there are no online transactions, the parking lot boasts license plates from around the South (830,000 visitors in 2010). And UBC—which perhaps fittingly abuts a cemetery—has become a bona fide tourist attraction, the type you learn about in brochures stocked in the lobby of the Days Inn up at Highway 35.

Inside, you'll encounter everything imaginable. As you roam the endless aisles, you'll find more ski boots than at Sports Authority, more cameras than at Best Buy, and more bras than at Victoria's Secret. There are crossbows and arrows. A full set of weights. A digital drum set. You can pick up a Balzac short story collection for a buck. Sure, occasionally there will be a quirky, newsworthy treasure: a 41-karat emerald, a full suit of armor, a 1934 French newspaper, a shofar, a hand-hammered cross. But UBC sold four thousand iPods last year, the supply of baby strollers and baggage wheelies is endless, and new wedding dresses arrive every day. The feeling I have gazing at a toddler's Scooby Doo clogs is not unlike viewing NTSB accident scene photos: Who were all these passengers? And why didn't someone want that oversized photo of Mickey Mantle back?

"The airlines get a bad rap," said Brenda Cantrell, UBC's

director of marketing, when we sat down just as Alabama was recovering from its worst snowfall in a decade. She acknowledged that many passengers feel the airlines don't care, but said she supports her supply chain: "I think they do the best that they can." In fairness, Cantrell also pointed out the high incidence of passenger fraud, since many dishonest airline customers do *not* want their bags returned; dirty socks be damned if an insurance claim for jewelry and electronics is approved instead. An old airline maxim: Every lost watch was a Rolex, every necklace was from Tiffany's.

As for baggage fees, Cantrell is no fan, since they are just about the worst thing that ever happened to the UBC business model: "It's definitely in decline because more people are not checking bags." But despite the long odds, imagine her satisfaction when a suitcase was pried open recently and inside was clothing still price tagged by UBC—come full circle.

As for how the airlines are doing, this is one aspect of customer service in which they are not completely at fault, because in November 2002 the industry happily abdicated responsibility for baggage screening to the federal government, and specifically to the newly formed Transportation Security Administration. For Free Market versus Government watchers, this was epic. The result? According to the U.S. Department of Transportation's monthly mishandled baggage reports, such filings soared after the TSA took over and air traffic started picking up again after 9/11. There were 3.84 mishandled bag reports per 1,000 passengers in 2002, and that number rose every year for six consecutive years, before peaking at 7.05 in 2007, just prior to the economic collapse of 2008 and the resulting drop in passengers.

One aspect the DOT stats do not fully capture is customer satisfaction, and how the airlines respond when passengers complain about what the DOT terms "lost, damaged, pilfered, and stolen" luggage. That's why it's worth noting that after the TSA started screening bags the DOT's monthly database of consumer complaints saw a marked spike in gripes generated by baggage handling. In fact, as a percentage of total grievances, complaints over luggage nearly tripled between the summers of 2002 and 2004.

I wrote about this topic after numerous arrests of TSA screeners for pilfering were reported in Detroit, Fort Lauderdale, Miami, New Orleans, New York City, and Philadelphia (and famous victims such as Chevy Chase and Joan Rivers made the news). By September 2004, the TSA finally addressed the mounting backlog of complaints by adjudicating more than 17,600 passenger filings, at about $110 each, for "property damaged or lost when their checked baggage was screened for explosives."

Now that more passengers are flying again, the number of mishandled bags is increasing again. SITA, a Geneva-based aviation communications organization, released Baggage Report 2011 and confirmed that an increase in passenger traffic has led to a 6 percent increase in mishandled baggage, with North America among the regions "most affected." However, there are important points to be made that highlight systemic problems with baggage handling.

"Baggage has never been a priority for airlines," says Scott Mueller, an expert on the topic. "If the airlines were held accountable, then at least they would refund the fee if the bag is mishandled." Mueller is a man with a passion for reuniting passengers and their baggage. He's a seventeen-year industry veteran who headed up baggage services for Midwest Airlines, and during one five-year stretch not a single passenger on his carrier filed a complaint over mishandled bags.

For one thing, the rate of checked bags has decreased since U.S. airlines began charging fees for this service, so this factor needs to be considered when comparing mishandled baggage rates across several years.

These DOT monthly statistics on mishandled baggage are problematic. For starters, it's a self-reporting system, and traditionally airlines have not posted an impressive track record under such programs. I know from firsthand experience that airlines can be "creative" with their flight delay reporting, and there is ample evidence that self-reporting safety issues to the FAA has not worked well. As Mueller explains, "The DOT does have the right to do a random audit. They *can* do that, but I don't think it's a high priority. You can easily fudge your numbers."

In addition, he notes the statistics are skewed because the DOT's monthly rankings are based on *passengers boarded*, not *baggage checked*. Mueller states: "The end results are based on the assumption that all 1,000 passengers that board an airplane checked a bag. Now if only 500 passengers out of 1,000 who board an aircraft actually checked a bag, then the airline's statistic of 4.5 bags mishandled per 1,000 passengers boarded would actually be double." What's more, there are four classifications of mishandled baggage—lost, damaged, delayed, and pilfered—but airlines are not required to break out the percentages on these subcategories, so consumers do not know if Airline A has a rampant problem with baggage break-ins while Airline B has a chronic issue with losing bags.

Mueller is equally critical of the TSA and the airlines. At Midwest, pilferage claims quadrupled after the government assumed responsibility for screening, and he has seen TSA employees stealing in Orlando. He explains that the TSA is the only entity authorized to open a passenger's checked bag, and in many cases such screening is performed in a private room with only two employees, a scenario conducive to theft.

Then there is the customer service component. Passenger rights advocate Kate Hanni says, "The way the airlines handle claims is they summarily reject them the first time through. It's a real racket. There's no pressure on them to make the system better for passengers." Mueller concurs, and says this "absolutely" happens every day: "There definitely are a lot of claims that fall on deaf ears."

However, a bigger issue is that most passengers are playing in a rigged game without having read the fine print about how to file, when to file, and where to file a claim when their bag and/or its contents is missing or damaged. "There are a lot of issues with how bag claims are filed," explains Mueller. "The airlines don't do a good job of explaining all this to their customers. People always say, 'How am I supposed to know that?' Well, you bear the burden."

Considering all the problems passengers encounter with checked luggage, it's worth noting that baggage agreements don't cover carry-ons left behind in the aircraft cabin; carriers are *not* required to ensure that the BlackBerry you tucked into the seatback pocket finds its way back to you. "The airline has no responsibility whatsoever," Mueller says. "It's considered lost-and-found. Chances are you won't see it again." He notes that his employees have found watches, cameras, laptops, and even $3,600 in cash in the cabins of empty airplanes—but not all airline and outsourced workers do the right thing. In fact, some do the absolute wrong thing, as was made clear by a news story that hit the wires in the summer of 2010. French police arrested a forty-seven-year-old Air France flight attendant named "Lucie R." and charged her with stealing thousands of dollars in cash and jewelry from passengers while they slept.

In March 2011, Congressman Michael Capuano, a Massachusetts Democrat, introduced legislation that would "require refunds [of fees] for baggage that is lost, damaged, or delayed." The *Congressional Record* shows that after ten minutes of debate, the "Capuano Amendment" was defeated on a voice vote. Interestingly, around that time I received an internal document prepared by the Air Transport Association, renamed Airlines for America

(A4A), addressing talking points on the Capuano Amendment. Here's the airline trade group's reasoning: "The amendment is unnecessary given [the] historically low mishandled bag rate and competing baggage handling and fee policies." Huh? Carriers are charging customers for a service they are not providing—prompt and safe delivery of your luggage. How is this affected by the overall mishandling record or "competing" fee policies? One month later, Secretary LaHood announced this policy had been adopted by the DOT, Congress be damned. But unfortunately, refunding fees won't improve airline baggage handling.

Declining Customer Concern in a "Service Industry"

The airlines claim that the number of passengers who are inconvenienced is quite small considering the millions carried. But beleaguered passenger Dave Carroll notes that such percentages don't mean much to those whom the system fails: "You have airline executives who quote statistics—but they don't seem to care about those on the margins of the statistics." He adds, "If there's no integrity in the policies, then it's open season."

Of course, few carriers are competing on customer service these days. Charlie Leocha of the Consumer Travel Alliance explains: "As the low-cost carriers and as the comparability made everyone more competitive, the first thing to go was differentiation in customer service. It's not only executive management—it runs through the fabrics of the companies. The managers, the gate agents, the flight attendants working without contracts. The front-line employees are under the most stress."

For airline passengers in recent years, customer service has gotten worse. That's not opinion—that's documented fact. According to statistics, there have been more mishandled bags (despite the added baggage fees), more consumer complaints, more congestion, and more passengers bumped off flights.

But other key elements of poor customer service can't be encapsulated in statistics, though they have been captured by dozens of polls, surveys, and rankings. In June 2011, for example, a *Consumer Reports* survey of fifteen thousand readers found a "low opinion of today's flying experience."

There is so much bad juju surrounding airline customer service that sometimes I need to step back and wonder if hyperbole is overtaking reality. Could so many pissed-off passengers possibly be wrong? Luckily, one of the best barometers I know happens to be a trusted friend and colleague. Linda Burbank was the ombudsman for *Consumer Reports Travel Letter* when I was editor, and a few years later she joined me at USAToday.com, where she continues to serve as a consumer travel advocate. I've seen the way she fights for readers who have been wronged by airlines and other travel companies. (On behalf of a *CRTL* reader, Burbank once secured a $29,833 refund from Royal Caribbean and Expedia, an unprecedented action from the laissez-faire and largely unregulated cruise industry.)

I sat with her in a café in San Francisco and asked her if airline service has really gotten worse. Burbank considered this and said: "I think I'm seeing fewer complaints, but I don't think it's because service is better. Everyone is so beaten down. We've all just become resigned to bad service. We're in the flip-flop generation."

It's worth noting that Burbank sees customers at their worst. As she explained, "When I hear from people they're already over-the-top angry. First, there was the original problem. Second, they have not gotten a suitable response. They don't feel they have been heard. It's not about the money—I have had people who are as angry about one hundred dollars as about three thousand dollars." With airlines, she's often able to secure discounts for future travel for her wronged readers, but cash refunds are much harder.

As for what passengers are complaining about, there's little surprise in learning that baggage fees top Burbank's list. Another trend line is how the airlines respond to "big problems" such as widespread weather delays. She noted that masses of people are angry, yet it seems no one is available to provide assistance—at the airport, in reservations, or even online.

In fact, many consumer complaints are often triggered by a flight delay. Burbank recalled years ago consulting an airline cabin seating expert who noted the correlation between delays and overall customer satisfaction: "If the flight was late, then the coffee was cold, the flight attendants were mean, and the seats were uncomfortable. You're already so harried when there's a delay." (By the way, in 2009 I examined flight times for USA Today.com and discovered that it now takes longer to fly on many routes throughout the United States because flight times have been padded by as much as a half hour.)

On the positive side, Burbank sees fewer complaints about Southwest and JetBlue. Perhaps not coincidentally, those are the only large U.S. carriers not charging for first checked bags (the second bag on Southwest is gratis as well). For the record, the DOT's consumer complaint rankings for 2010 found Southwest ranked first among the nation's eighteen largest carriers, while JetBlue ranked thirteenth, demonstrating it has a long way to go in emulating Southwest. In addition, since the DOT began keeping records in 1987, Southwest has led all other domestic carriers with the best on-time performance record and the fewest passenger complaints.

Suing an Airline Is Usually Not an Option

The general rule is that you can't sue a U.S. airline in state or local courts because the airline business is one of a handful of industries that have been given federal preemption. However, there are exceptions. The DOT's Aviation Consumer Protection Division even provides an online manual titled "Tell It to the Judge: A Consumer's Guide to Small Claims Courts." The DOT cites examples of situations in which you might consider using such a court; of course, in all cases the key factor is the amount of the claim being filed, since there are limits on monetary damages.

Attorney Al Anolik also recommends this more proactive course for passengers who have been wronged by domestic airlines. In fact, a large part of his practice involves litigating such cases, often handling ten or twelve cases in one morning. Anolik explains that small claims courts have more liberal rules for admitting cases, which can work in consumers' favor. What's more, airlines have a duty to mitigate, so they can't ignore such filings.

Anolik clearly has done okay by leading such fights as one of the nation's most knowledgeable travel attorneys. His beautiful home offers a spectacular panorama of San Francisco Bay that encompasses both the Golden Gate and Bay bridges, a view shared by neighbors Andre Agassi and Steffi Graf. I asked him if I am wrong in concluding the major airlines have gone out of their way to confuse customers by intentionally muddying their contracts of carriage, those vital documents that spell out exactly what passengers can expect. Anolik laughed and said, "The contracts of carriage definitely have been tightened up. Very little in the contracts gives you a right, but they will mention the rights they're taking away." He added, "Every time the airlines get hit with a lawsuit they fill in a loophole. If there is a new cockamamie charge and it is not on the contract of carriage, they put it in." What's more, Anolik noted that these exemption lists keep growing, even though the DOT ordered a moratorium on such exemptions. His blunt assessment: "It's fraud."

These contracts spell out what an airline will do if you are bumped, or your flight is delayed or canceled. And what we found after studying the issue at *Consumer Reports* is that over the last decade these documents have become murkier, as plain English has been replaced with legalese. And a term such as "the airline shall" has been supplanted by "the airline may" instead.

In theory, U.S. airlines are supposed to prominently post these contracts on their websites, but good luck in finding most of them in less than five page clicks. (A notable exception is JetBlue, which has provided the most customer service transparency in the industry since its Valentine's Day weekend operational meltdown in 2007, when 1,100 flights were canceled.) I have personally found that airline call center employees and even airport personnel usually have no idea what these contracts are, let alone what they dictate.

Not that clear explanations will help. "It's really a unilateral contract, not a bilateral contract," says passenger advocate Kate Hanni. And sometimes the legalese can be just plain misleading. In July 2010 a Southwest Airlines executive corrected a "misinterpretation" over the term "mechanical difficulties" being included in its (long) list of force majeure ("act of God") flight delay conditions; he explained the "added verbiage" referred only to mechanical difficulties outside the airline's control, such as a broken airport deicing system, and not aircraft problems. The incident underscored just how confusing such "added verbiage" can be for passengers.

Then again, many of these "employees" are not employees at all, a topic I discussed with Anolik. And he clearly relishes torturing the industry: "When I'm early for a flight I'll ask, 'Can I see your denied-boarding policy?' And the person working will have no idea. You're right, they don't know the rules anymore."

Airline industry old-timers—both employees and passengers—still invoke the days of the Rule 240 clause, a holdover from the regulated era in which the government spelled out specifically how airlines were required to meet their passengers' needs. It was clear and concise for all passengers and uniformly fair for all carriers, but Anolik said Rule 240 is no longer mandatory in the deregulated era.

However, I believe it's time to write a new Rule 240, for the twenty-first century. That's why when I became a member of the FAAC I urged the DOT to adopt procedures similar to those employed by the European Union. The EU's rules are clear, cogent, and easy for every passenger to understand, particularly if they download a color-coded chart that indicates uniform compensation for airline mistreatment. This isn't to suggest that the European model should be copied outright, but it certainly provides a decent blueprint for us. As travel ombudsman Linda Burbank notes, "They have better transparency in Europe. In America it's about flying Darwin Air."

As a last recourse, passengers can still file a formal complaint against a domestic airline with the DOT. Though many consumers undoubtedly respond with a "What's the point?" attitude about such a pro forma task, Anolik noted that it's really quite important: "Unless they hear from passengers, the DOT will say, 'We don't have any complaints on file.'"

Loads of Fun? Airlines Emulate Troop Carriers

One mechanic for a legacy carrier—so-called because these large airlines predated deregulation—summed it up: "If you're flying full planes and you can't make money, you shouldn't run an airline." And yet that's exactly what we have. My own theory is that the airline industry's decision to fill all these planes has had a direct effect on making flying more miserable: more boarding delays, more mishandled bags, more consumer complaints, more air rage.

During World War II, when commercial airlines were pressed into service as de facto military transports, planes were fuller than at any time before or since, with average passenger load factors—the percentage of occupied seats—reaching nearly 90 percent. After 1946, it took the U.S. airline industry more than fifty years to crack the 70 percent mark again. By 2009, load factors reached 80 percent, and by 2010 the domestic industry's average topped out at 82.1 percent. With an average that high, the percentage of flights at or near 100 percent full obviously has increased as well. But this is not what airline executives ever envisioned.

Analyst Bob Harrell explains: "With airlines, everybody's in that game to get the last twenty passengers in the bucket. But it creates a terrible service problem. Transportation systems are designed for 65 percent to 70 percent capacity and they can't handle it when it's twenty points higher. Which you'll see if you're in the middle seat of a 757 waiting to get served."

Consider it this way. Delta operates Boeing 737-800s configured in economy class with three seats on either side of the aisle for a total of 144 seats, or 6 in each of the 24 rows. Putting aside those passengers who are traveling together, leaving all 48 middle seats empty in this 3x3 configuration would require a 67 percent load factor. An 80 percent load factor means only 29 middle seats will remain unoccupied; a 90 percent factor leaves only 14 empty middles. Further, consider that every middle seat taken causes discomfort for not one but three people. Obviously that's a lot of crowded passengers, a lot of overstuffed overhead bins, a lot of squeezing and jostling in the aisles, a lot of waiting for that lavatory.

There's a technical term economists employ to describe this condition: it's called *greed.* Over the last few years the same airline executives who have levied fees for checking bags and calling reservations, who have outsourced flying to low-cost regional airlines and maintenance to Third World sweat shops, have decided that the airline business most closely resembles the sardine canning industry.

What's more, industry experts note that not only are all those full planes not good for passengers, they're not even good for the airlines themselves. That's because once an airline's loads hit the 70 percent mark, that carrier starts "feeding its competitors" because it can't handle the spillover. In addition, higher loads create additional operating expenses.

"It's awful," said Rolfe Shellenberger, an industry legend who began his thirty-one-year career for American Airlines as a reservations agent in 1951. "They're cattle cars now." I shared with him my contention that packed airplanes have eroded customer service, and he agreed, saying full planes have greatly contributed to this deterioration.

Shellenberger also pointed out what he calls a "curious anomaly"—that Southwest has the lowest load factors among the majors, yet has been the most consistently profitable. And it's true: in 2010, Southwest's load factor was 79.3 percent, lower than the domestic load factors for American (82.6 percent), Delta (82.9 percent), United/Continental (84.9 percent), and US Airways (83.2 percent).

And there's yet another negative side effect to all those crowded airplanes. In early 2011, the aviation consulting firm Oliver Wyman released its Airline Economic Analysis and noted: "In a sense, as load factor has approached its theoretical maximum, fees have replaced it as a source of revenue growth." That's right: as long as cabins remain full, the big airlines will continue ratcheting up those fees.

Things That Go Bump: Airlines Deny Boarding

Those record-high load factors are fueling another disturbing trend: more passengers being bumped against their will. The industry calls them involuntary denied boardings and the DOT calls them oversales, but by any term it means your airline seat has been given to someone else without your consent. And it's a practice that has become more prevalent in recent years. According to the DOT, 1.09 passengers were denied seats for every 10,000 passengers boarded in 2010, an increase over the 0.89 rate in 2005.

As Forbes.com noted, Broadway theaters don't double-book seats: "No other (legitimate) businesses sell the same product to more than one customer, so why do airlines?" It's an excellent question, but only industry apologists defend the practice, citing the vague complexities of yield management practices so vital to the airlines' elusive profitability. They also note that passengers with refundable tickets often no-show for flights, thus leaving the airlines with empty, unpaid seats.

The good news is that the DOT requires that airlines provide those bumped against their will with denied boarding compensation, and in 2011 Secretary LaHood announced the DOT had doubled the limits on these amounts.

Meat in the Seats

Industry executives have had an attitude about airline seating for generations. In fact, designers once eagerly met with Eddie Rickenbacker, the famed World War I flying ace who became CEO of Eastern Air Lines, to show off their newest seat cover fabrics for the Lockheed Electra. Rickenbacker's legendary response: "I don't care what you cover the seats with as long as you cover them with assholes." Boy, have the latest crop of airline execs gotten good at that.

Even Ralph Nader, who arguably knows more about customer satisfaction than anyone in America, sums up the airline industry in this way: "It's about the consumer mistreatment. You need to buy one seat for your head and torso and then another seat for your knees."

Obviously so much of the flying experience comes down to legroom. In fact, when it comes to airline seating, it's specifically about something called *pitch*, an industry term for the front-to-back measurement of the distance between seats. SeatGuru.com is completely devoted to airline seating, so I asked Jami Counter of TripAdvisor, the site's parent company, about the big squeeze. He agreed that full airplanes are affecting passenger contentment, and not in a good way: the more crowded the plane, the less comfortable the seats. "Load factors, that's the biggest factor," he said. "Overall, that leads to a negative perception for passengers." He added that most travelers will suck it up on short flights of two hours or less, but expect more comfort on longer journeys: "At SeatGuru, our sweet spot is the medium- to long-haul flights. The three-to-four-hour market."

I wonder if seats have shrunk in recent years, but Counter advised that, for the most part, U.S. domestic carriers have "stayed constant" with economy class seat pitch of about 31 to 32 inches on average. So which domestic airline earns kudos? It's no contest, since one carrier provides seat pitch of 34 to 38 inches in economy: "JetBlue is the best by far." (This was confirmed by *Consumer Reports* last year, when our airline survey ranked JetBlue at the top for seating comfort, and eight of ten carriers received low scores in that category.) And which is worst? Once again, there are no serious competitors since one airline is offering just 28-inch pitch: Spirit. Counter said, "That's just cruel."

But legroom is only half the battle—there's also seat width to consider. Counter explained that evaluating width makes for an apples-to-oranges comparison, because some airlines offer slimline seats, newer lightweight models that offer more room. He also contended that the manner in which seats recline affects passenger comfort as well.

In 2002, *Consumer Reports Travel Letter*'s annual review of the best and worst airline seats included a report on a British ergonomics firm that found most standard airline seats are "totally inadequate" for larger passengers. While seat width generally falls in the range of 17 to 18 inches, an anthropometric table comparing average butt sizes worldwide indicated America ranked first—no surprise there—but at 20.6 inches, that's a pretty tight squeeze in most economy seats. And there certainly is no indication that American butts have shrunk in the last decade.

To be fair to Boeing and Airbus and other aircraft manufacturers, onboard comfort usually has very little to do with the airplane itself, and everything to do with the airline configuring it. If you peel back the carpeting on a commercial jet, you'll find that most passenger seats are locked into tracks or pallets and therefore the distances can be adjusted, between seats as well as between rows. For airline executives, of course, it's all about cramming in more bodies—and they've all but perfected that dark art. Back in 1999 I wrote an article for *New York* magazine titled "Sky Box: Shag or Shul?" which detailed how customized cargo containers could be used by Virgin Atlantic to provide lounges, showers, and exercise and massage areas on its new Airbus fleet, while El Al could employ the same equipment as airborne temples on its Boeing planes. Neither of these plans was fully implemented, of course, because in the end airlines are all about carrying more stuff—be it cargo, mail, or people—and not about wasting precious space. So much for those Pan Am 747 piano bars that look so cool in the retro ads.

Labor leader Pat Friend, who began her career as a flight attendant in 1966, points out that United's Boeing 757s used to be configured with a coat closet up front, but a few years ago it was replaced with another row of seats. When flight attendants asked management about it, they were told, "You can't sell a coat closet."

Now that the sacred cow of ancillary revenue has been slaughtered, amenities such as better seats and upgrades come with a price tag. On the Airfarewatchdog site, George Hobica recently responded to readers seeking the secret to nabbing an exit row seat: "And if you have to ask, you probably aren't going to get one. Not without spending some money, anyway."

What's certain is that airlines will continue wedging in as many seats as possible, particularly in economy. So that requires that passengers not only work harder at selecting preferred seats at the time of booking, but also wade in with elbows and knees when it's time to board the airplane. But all that jostling and wrestling raises another issue: in-flight etiquette.

"Civil" Aviation? Responsibility for Onboard Rage

If there was a demarcation line in the ongoing battle between passengers and airline employees, it undoubtedly came on August 9, 2010, when JetBlue Flight 1052 arrived at JFK in New York City. The details blur on the actions or nonactions of an aggressive passenger, but all agree that harried flight attendant Steven Slater spewed obscenities into the PA system, opened a door still armed as an evacuation slide, and bid adieu to a twenty-year career in aviation with beer can in hand. What followed was an American rite of passage. Arrest. Arraignment. Media storm. Morning talk shows. Folk hero to some, unhinged alcoholic to others.

Nearly a year after his day of infamy, I had a long and engaging conversation with Slater. I found him funny and insightful as we swapped industry war stories, and he told me he still loves aviation. He's from an airline family—his father was a pilot for American and his mother was a flight attendant—and he began working for a string of airlines in 1990. (In fact, as a TWA flight attendant, in July 1996 he was in Rome, waiting to return to the United States on the 747 that exploded over Long Island as Flight 800.) Eventually, Slater landed at JetBlue, but he told me that the fun of working for the low-cost carrier wore off quickly.

"One huge resentment I have is that I made less in my twentieth year in the airlines than I did in my first year," Slater said. He continued flying because he needed JetBlue's health benefits and flight privileges, but by 2010 he was making $9,700 annually and commuting from his New York home to nurse his dying mother in California. Meanwhile, he was angered by what he terms JetBlue's "lean" operation: "It feels like you're being taken advantage of when you see new paint schemes and a new headquarters and the company naming sports stadiums." At the same time, he says, the company had no money to improve conditions for crews.

I told him that I sympathized, that I certainly understood how airline employees, particularly flight attendants, have been in the front lines as passengers revolt against crowded flights and packed overhead bins. But I couldn't justify "popping the slide," since like all former airline employees I know that opening a door armed by compressed carbon dioxide and nitrogen could kill an innocent bystander. But Slater maintained that he assessed the situation, and told me, "I absolutely acknowledge the inherent danger. I know how to assess conditions. It's not like I shot a gun in a crowded movie theater." Then he added, "If anyone knows how to open a frigging door, it's me."

But what no one, including Slater, could have foreseen was how quickly his story would resonate—not just with fed-up employees at airlines, but with fed-up employees everywhere. He acknowledged he is always going to be "a very polarizing and divisive figure" but said he received emails from thousands of supporters, particularly from nurses, police officers, and firefighters. So I asked Slater a simple question: What drove your popularity?

"It was the first time someone stood up and said enough," he told me. "The thing is, *everybody* is overworked and underpaid and outsourced." He also summed up how he views it now: "I don't use the word *regret*—it's wasted energy. It might not have been the most thought-out route of egress. But I sure felt a hell of a lot better."

Fans and foes alike all concur on one point: Slater's take-this-job-and-shove-it episode touched a national nerve. As columnist Peggy Noonan noted immediately afterward, "Once we were a great industrial nation. Now we are a service economy. Which means we are forced to interact with each other, every day, in person and by phone and email. And it's making us all a little mad."

What Would Emily Post Do?

If Steven Slater's actions were anomalous, that would provide a certain context. If he were the only person to have lost it on a commercial jet in recent years, it would be a different story. However, a quick Google search indicates that a week doesn't go by without some type of air rage–induced drama playing out in the troposphere. Sometimes the headlines tell the tale:

- "Intoxicated Playboy Models Arrested After Rowdy In-Flight Antics"
- "Smoking Woman in Air Rage"
- "FBI Busts Rabbi for In-Flight Groping"
- "Brothers Accused of Beating Airline Pilot"
- "Getting Off in Denver: Programmer Arrested for In-Flight Masturbation"
- "Flier Blames Tabasco Spill for Lewd Act"
- "Gerard Depardieu Urinates on Plane"

But if we're all so unstable, why don't we hear of Greyhound rage or Amtrak rage? In other words, as with many types of crime, are systemic factors at play? Do the airlines actually *induce* bad behavior? Or as transportation strategist Stephen Van Beek has put it, "In the airline industry, chivalry is completely dead. I've seen older passengers pushed aside." The airlines may not consciously *invite* bad behavior aloft, but most carriers certainly do nothing to address the two biggest contributing factors: the record passenger loads and the lack of adequate overhead bin space now that checked baggage comes with fees.

And small gestures could help. One former flight attendant gives credit to US Airways for routinely avoiding cabin dustups by properly screening passengers *before* they tote oversized bags on board, unlike many airlines that regularly experience delays while large bags are tagged in the aisles.

But in truth the hassles begin for most passengers back in the airport, particularly once the TSA begins its groping and poking and radiating. That's why veteran airline executive Howard Putnam believes the hassle factor is beyond the industry's control to an extent: "It starts when you enter the door of the terminal. You can't dump your wrath on the TSA, or you'll go to jail. So you dump your wrath on the airline."

It's a fair question to ask: are we also to blame for our own miserable in-flight experiences? The Australian site News.com.au recently detailed "What Flight Attendants Hate About Passengers," which included the obvious—dealing with drunks. But they also cited passengers not paying attention during the safety demo; switching seats without regard to weight and balance considerations; dirtying lavatories (termed "overshooting the runway"); and not making eye contact or saying "thank you." Conversely, two years earlier passengers had sounded off about "hosties" with complaints that included flirting with male passengers, bumping against those in aisle seats, and disappearing when the call button is activated. Interestingly, a disconnect was highlighted when passengers complained about flight attendants not helping them stow luggage in overhead bins—and flight attendants complained about even being asked for such assistance. And some passengers were annoyed about being told to turn off their electronics.

Obviously these conditions require all of us to redouble our efforts at being civil (especially in the lavatories). Though Emily Post is no longer around, luckily her great-granddaughter-in-law Peggy Post carries on her mannered legacy as an etiquette expert, author, and director of The Emily Post Institute. And, no, Peggy Post didn't spend her young adulthood being quizzed on salad forks at finishing school. In fact, she had the best possible training for a life spent examining human behavior: for two years she traveled the world as a flight attendant for Pan Am. "That's a much tougher job now," she says. "You're in the front lines of etiquette."

I've spoken to Post on several occasions, though I'll admit I was fairly nervous the first time; is "good day" preferable to "good afternoon"? But her relaxed manner immediately puts one at ease, especially when we swap Pan Am stories. Like every other veteran airline employee I've chatted with in recent years, Post agreed it's become a different industry—especially with all those crowded airplanes.

"In some ways the airlines are doing better and passengers are doing worse," she said about the high load factors. "It's a perfect storm." The combination of crowded planes, a lack of meal service, and a fear of flying can cause some passengers to "lash out" at airline employees and each other. Post noted certain hot-button issues that commonly lead to onboard kerfuffles, including the battle for precious overhead bin space, the "to-recline-or-not-to-recline-the-seat" debate, the unwrapping of brown-bagged "smelly food," and the mad rush to deplane.

Even so, she doesn't believe that airline policies and other systemic causes provide license for insolence: "Let's not forget there are some really, really rude people out there." But as someone who once pushed a meal cart down an aircraft aisle, she knows from which she speaks when she advises airline employees: "There's never an excuse for not being civil."

Your Call Is Important to Us

As for the O-word—*outsourcing*—there's no doubt it's had an ill effect on baggage claims as well. In January 2011, Jaunted.com featured a long and detailed article, "The Incandescent Incompetence of US Airways' Outsourced Baggage Recovery Call Center," which included this summation: "The short version is that US Airways has an outsourced baggage recovery call center, almost certainly in India, that's either unable or unwilling to give customers reliable information or genuine assistance." Ouch.

Occasionally, however, a high-profile case of airline indifference focuses widespread attention on an uncaring industry. Enter Dave Carroll, the Canadian musician whose Taylor 710 acoustic guitar was manhandled at O'Hare International Airport when he was traveling from Halifax through Chicago to a gig in Omaha in March 2008. As he and his bandmates were settling into their seats on the connecting flight, a female passenger watching the baggage being loaded suddenly exclaimed, "Oh my God, they're throwing guitars outside!" The worst was confirmed in Omaha: Carroll's 710 was badly damaged and written off—but his odyssey was just beginning.

What followed for Carroll was standard operating procedure for millions of other airline passengers, the proverbial runaround from the airline (in this case United) and what he terms "widespread indifference." There were long pauses between communications; his calls and emails were seemingly ignored; he received a letter from the airline with no name or return address; he sat on hold as his call was transferred to India. United was far from contrite. He should have filed the claim within twenty-four hours. There was nothing they could do. It was regretful. Ibid.

"If I was a lawyer I would have sued," said Carroll. "But instead I wrote a song." And what a song. Last year "United Breaks Guitars" passed eleven million views on YouTube and had been heard on U.S. and Canadian television. In fact, I first encountered Carroll in 2009, when he performed the unplugged version in the halls of the Rayburn House Office Building at a forum on passenger rights issues.

Now in most cases, it would be fair to assume Carroll's fifteen minutes have long expired, particularly after he followed up his opus with two more ballads in the United trilogy. But his story transcends his one-hit-wonder status and clearly taps into a wellspring of frustration and anger at how airlines handle—or rather do *not* handle—complaints.

"I was really frustrated and I wasn't about to let it end," he explained. "I was trying to move an immovable object and when you do that you only hurt yourself. I was throwing myself at a brick wall—but what if this brick wall was sitting on a shaky foundation?" In fact, the most remarkable aspect of Carroll's story is how shabbily United reacted even *after* he had become a bona fide North American celebrity. "It was handled so poorly," he said. "It should be what we learned in kindergarten about right and wrong."

This is no doubt why many question if the dysfunction is intentionally built into airline customer service programs. In other words, the airlines do *not* want to hear our gripes and have constructed a byzantine system designed to thwart us from pursuing our ultimate goal of customer satisfaction. It's not unlike the panopticon, the eighteenth-century prison design for a structure in which guards can view the inmates even though the incarcerated are unable to tell if they themselves are being watched.

Why else would they remove human beings from the airports and the call centers and ask you to write instead? Why else would they outsource complaints to Southeast Asia, knowing that cultural differences could impede your satisfaction? Why else would they impose fees if you choose to speak to a human being? Why else would United shut down its telephone complaint center and replace it with an email model? (A spokeswoman was quoted thus: "We did a lot of research, we looked into it, and people who email or write us are more satisfied with our responses.") Why else would AlaskaAir.com channel distant memories of artificial intelligence icon Max Headroom by introducing "Jenn, Your Virtual Assistant"?

A world in which all airlines wish to respond quickly to their customers' concerns would have no place for GetHuman.com, an ingenious and at times invaluable site that provides shortcuts for those wishing to speak to an actual person after dialing a toll-free number. For example, here's the simple but secret way to bypass Delta's customer service Maginot Line: "Press 0 at each prompt, ignoring messages." What's more, GetHuman.com provides typical wait times and user ratings as well (13.8 minutes for Delta, which rates an "average").

Stuck on the Tarmac with You

In the fall of 2009 I took part in an airline industry forum in Washington, D.C., hosted by the DOT. I spoke about a troubling array of issues affecting airline passengers. At that forum Spirit

Airlines CEO Ben Baldanza and Republic Airways CEO Bryan Bedford spoke of their confusion over the need for a Passenger Bill of Rights and asserted that passengers could always "vote with their feet." I responded that I had done just that, by taking Amtrak to Washington that morning. Not everyone in the room was amused.

The following summer, at the second full meeting of the FAAC, I read a statement into the record to address the "vote with their feet" issue, as if such a thing were possible at thirty-five thousand feet. I noted that the nation's aviation infrastructure belongs to its citizens and taxpayers, and that in many markets there is no meaningful choice, through either lack of competition or airlines refusing to compete. What's more, the federal preemption rule and the industry's no-nonsense safety and security regulations don't allow for much consumer debate.

I had long been a journalist fighting for passengers, but by late 2009 my work for Consumers Union was pulling me further in the direction of passenger advocacy. I soon found it was becoming a rather crowded field.

Kate Hanni had no legal, political, or aviation experience when she became a national symbol of the passenger rights movement. She was an American Airlines passenger who endured a horrific fifty-seven-hour ordeal, with nine of those hours spent on a tarmac in Austin, Texas, alongside her husband and two children in 2006. There was no water or food, the lavatories overflowed, passengers became sick, and Hanni felt trapped—again. Six months earlier she had suffered an attempted rape and murder and, Hanni explained, "In a way I was the worst person to do this to. I was not going to be a victim again."

She went on to establish what evolved into FlyersRights.org and soon became a ubiquitous presence in the media. And she deserves credit for lobbying DOT secretary LaHood to eventually impose the "three-hour rule" in 2009, which requires that planes operated by major airlines return to the gate within that time frame if departure is not imminent (the DOT later expanded the rule to include smaller carriers and foreign airlines).

But there are those who feel the government—and particularly LaHood's DOT—is overreaching through such actions. Brett Snyder is an airline veteran who spent time at America West and United before launching an online discussion forum while working for PriceGrabber.com. Eventually he spun it off into The Cranky Flier blog, where "the guiding principle is what I find interesting." Much of Snyder's focus recently has been on customer service, but he doesn't echo the legion of passenger rights advocates—in fact, he often criticizes them—I mean, us. He explains: "The government feels the need to step in on the airline industry more so than in other industries. Part of it is a legacy thing. . . . What I object to are the things that will create more problems than they will solve, like tarmac delays. With the three-hour tarmac rule, you'll have a greater number of people inconvenienced than not. That's my concern."

A few years ago I might have agreed with him. In fact, I took quite a few lumps for being the only passenger advocate who lobbied *against* tarmac delay legislation back in 2007. In my USA Today.com column that year I wrote: "It's tough to decide how best to fix airline customer service. That's why I have such mixed feelings about Congress micro-managing flight operations. The journalist who fights for passenger rights has been waging an internal battle with the ex-dispatcher who knows that even large doses of outside assistance will not necessarily correct systemic airline operations problems."

But my fence straddling ended in September 2009, when I attended a passenger forum cosponsored by Senator Barbara Boxer of California. That morning the public heard from dozens of passengers who had encountered lengthy and inexcusable tarmac delays, as well as from passenger rights advocates and legislators. But not a single sitting airline executive attended, and even the industry's primary lobbying group—the Air Transport Association—declined the invite.

That day it became apparent to me that if the airlines refused even to *listen* to passengers, then they deserved whatever regulation and/or legislation Washington deemed appropriate. Even those who oppose the three-hour mandate acknowledge that the airlines—both individually and collectively—repeatedly muffed the chance to promote an alternative policy, *any* alternative policy. Analyst Bob Mann, who maintains the DOT rule could inconvenience tens of thousands of passengers, also declares, "The airlines were derelict, no question."

Michael Levine, a law professor and former airline executive, is a strong proponent of free-market solutions, but even he acknowledges this dichotomy: "I think the airlines have an odd mixture of political sophistication and naïveté. They certainly have an extensive lobbying effort and they know how to deploy their executives on Capitol Hill on issues. But often their choice of issues and their response to issues just doesn't reflect what anyone who would step back from the politics would suggest they do."

As someone who once was responsible for creating, delaying, consolidating, and canceling dozens of flights a day as a flight operations manager, I can tell you that not all passenger needs are the same; when a long delay is announced, some will want to leave the aircraft or perhaps even leave the airport, while others will want to wait it out on board and hope for the best. In addition, the destination and flight length are critical components. At the Pan Am Shuttle on days when LaGuardia was particularly backed up, we would board passengers at the gate and then drop the aft staircase embedded into the Boeing 727 and allow customers to leave at will. (That same 727 aft staircase, by the way, is what allowed famed hijacker D. B. Cooper to parachute out over

Washington state with his loot.) The new tarmac regulations are not perfect, but they do provide enough flexibility to accommodate both types of passengers.

"Nobody should be required to sit on an airplane for more than three hours," says former American Airlines CEO Bob Crandall. "From a public relations point of view, [the airlines] were completely tone-deaf. Moreover, the airlines and the airports have resisted and those problems can be easily resolved. This whole nonsense about canceling flights. Bullshit. I'm not going to cancel a flight. I've been in this conga line now for two and a half hours. . . . Anybody wants to get off, get off. End of discussion."

Where I do agree with Snyder is that the focus on tarmac delays has seemed to overshadow much larger passenger issues. In 2010, shortly after the DOT enacted the new rules, the head of a large domestic airline said to me off the record: "You got your tarmac delay rules, what do you want now?" That's not to detract from the experiences of those who have been trapped on airplanes and not provided water, food, and lavatories for hours on end. But those passengers make up an infinitesimally small statistic; meanwhile, the rights of tens of thousands of passengers are abused every week owing to "routine" overbookings and flight delays and cancellations. Snyder says, "I agree. There are very few people impacted by tarmac delays. And a lot more are impacted by other factors every day."

"We Owe Him Nothing"

Some carriers actually seem to embrace their public disregard for passengers. Take Spirit Airlines. A *USA Today* profile in 2009 quoted Ben Baldanza referring to his company as "the McDonald's of the airline industry." It also referenced the infamous tale of Baldanza accidentally responding directly to a customer's email requesting a refund for a flight delay; Baldanza had written: "Please respond, Pasquale, but we owe him nothing as far as I'm concerned. Let him tell the world how bad we are. He's never flown us before anyway and will be back when we save him a penny." Meanwhile, the DOT repeatedly levies fines against Spirit for false advertising and failing to provide bottom-line pricing inclusive of fees.

An employee at Spirit recently told me about a warning from a Spirit veteran: "If you want to succeed here you're going to have to forget everything you learned at other airlines." My friends Larry Bleidner and Irene Zutell are frequent flyers between their home in Los Angeles and their families on the East Coast, but even they were shocked when they attempted to save a few bucks on Spirit. Although they encountered five-hour delays in both directions, they were never provided updates, explanations, or apologies, and they watched as their delayed flight simply rolled off the information display screen near the gate as the pilot shrugged. Bleidner says, "The misery index was like on a New York City subway stuck in a tunnel—but that's back when it was a buck. You'd expect at least a modicum of respect from an airline." On another journey, Zutell was horrified to spend five hours on the tarmac with her young daughter while an inoperative lavatory was fixed: "They stuck us on a plane for five hours so they could fix a bathroom for a two-hour flight. Even my five-year-old knew it didn't make sense."

I hear such airline war stories repeatedly, so I'm always questioning my perspective. I met up in Washington with Charlie Leocha, director of the Consumer Travel Alliance, and asked his take on the big picture. He told me: "I think we've moved from an industry where all competition was based on customer service. Once the industry was deregulated, the real competition became price competition. Then two things happened. First, price became more important. Second, airline executives no longer were aviation people. They became MBAs. And passengers were no longer passengers, they became statistics." He underscored how important LaHood has been: "Finally the DOT is helping us out. It's critical. Because we have no standing in the courts due to federal preemption. Therefore we have the rights of medieval serfs because we cannot petition our noble lords and masters at the airlines. So we petition the DOT to intercede on our behalf."

In the end, Raymond LaHood, a former congressman and one of two Republicans in President Obama's cabinet, did what neither the House nor the Senate could do: he strengthened passenger rights. First came the three-hour tarmac delay rule in 2010. And then in April 2011 he announced a new set of provisions:

- greater transparency in fares and fees
- refunds of baggage fees if the bags are lost
- increased compensation for involuntary bumping
- a prohibition on raising fares after tickets are purchased
- an expansion of the tarmac delay rules

I asked LaHood to summarize, which he did thus: "I served in Congress fourteen years and Congress never passed one bill having to do with passenger rights. We were here for two years and we got it done. So you can cross your fingers and all your toes and your legs and arms and everything else with the hope that Congress will pass legislation. But I'm not optimistic about that. We have done what Congress couldn't do for a decade, and we did it in two years."

My concern, one shared by many others, is that the DOT simply has not done enough to rein in an industry that clearly wants to police itself. Furthermore, through I would give LaHood an A for his work on passenger rights, the DOT and its subsidiary the FAA seem incapable of providing effective oversight of aircraft maintenance, outsourcing, regional carriers, pilot proficiency, infant restraints, and other life-and-death safety issues.

2

What Happened to the Airlines?

Capitalism without bankruptcy is like Christianity without hell.
—*Frank Borman, astronaut and CEO of Eastern Air Lines*

AT THE VERY FIRST MEETING of the DOT's Future of Aviation Advisory Committee, Raymond LaHood instructed us that two topics were off the table. One was aviation security, since that falls under the purview of the Department of Homeland Security. The other was reregulation of the airline industry, which would not be considered by the Obama administration. Yet in some ways, the battle over airline deregulation reflects the larger schism occurring in American politics. There are two warring camps, and each side wonders how the other could possibly be so blind—and so wrong.

Proponents of deregulation claim that nearly three times as many Americans are flying, and when adjusted for inflation, airfares are lower than they've been in decades, while safety has improved. Critics maintain that service has deteriorated, airplane cabins are overstuffed, labor relations have imploded, airline balance sheets are a wreck, and outsourcing is compromising quality, service, and even safety. To varying degrees, there is truth in both arguments. What's more, at times it is difficult to determine what effects, both good and bad, are directly attributable to deregulation itself, and which changes would have occurred anyway over three decades. Yet in my view, we've reached a tipping point, and the bad effects are outweighing the good.

"Clearly consumers are saving money in the deregulated era," former congressman James Oberstar says. He points to more nonstop service and fewer one-stop flights. "But there are customer service issues and maintenance outsourcing concerns as well as concerns about consolidation." Coming to terms with those concerns could be the most contentious battle the airline industry has ever endured.

When Flying Was Fun (1903–78)

It may be hard for younger people to understand that flying was not experienced by most Americans prior to the 1970s. (My father spent four years in the U.S. Army in World War II, and served on three continents before being wounded at Anzio, but all his travels were by truck, train, and transport ship.) A full sixty years after Kitty Hawk, only 15–20 percent of American adults had ever flown on an airline. Today those numbers have been reversed, with about 85 percent of the population having flown. Ed Perkins, my predecessor as editor of *Consumer Reports Travel Letter*, believes there's been a societal shift: "What's happened with the airline industry has happened in so many other industries." He notes it started primarily for the wealthy, but because of technological advances aviation spread to the masses. In fact, Perkins compares flying to skiing, which was once an upper-class pursuit but has trickled down to virtually all classes.

Therefore, any discussion of the U.S. airline industry is cleaved by the watershed year of 1978, the way American history hinges on historic dates such as 1775, 1865, or 1945. Because in

1978 the airline business for the first time became just that—a business. And nothing has been the same since.

Starting in 1937, the Civil Aeronautics Board had treated interstate commercial airlines as a utility and determined which carriers operated on which routes, and on which days and at what times. The CAB also decided how much airlines could charge for fares, and ensured they received a return on investment. Interestingly, carriers that operated within the borders of a single state were free from regulation, such as Pacific Southwest Airlines in California and Southwest Airlines in Texas.

The process was cumbersome, and old-timers joke that Washington bureaucrats threw darts at a board to determine the fares between Cleveland and Minneapolis. Airlines that wanted to expand their route maps were subject to interminable red tape and could wait years for responses. There was an air of gentility about the airlines, and most U.S. carriers were restricted to domestic routes while Pan Am—and to a lesser extent TWA and Eastern—flew the American flag overseas.

Then President Gerald Ford, followed by President Jimmy Carter, expressed support for deregulating commercial aviation, and the movement came to be seen as a precursor of deregulation in other industries, such as telecommunications and trucking. The man chosen to usher in the future was Dr. Alfred Kahn of Cornell University, who had written *The Economics of Deregulation*, the seminal work on the topic. Kahn was tapped by Carter to head up the CAB—and then dismantle it.

Perhaps one anecdote sums it up. When PBS broadcast an examination of airline deregulation in 2002, Associate Justice of the Supreme Court Stephen Breyer was among those interviewed, because back in the 1970s he was chief counsel for the U.S. Senate Committee on the Judiciary. In that role he worked closely with Senator Ted Kennedy in shaping the Airline Deregulation Act, which was signed into law on October 24, 1978.

On PBS, Breyer recalled how Lamar Mews, then the president of Southwest Airlines, attended a hearing and said: "The people who put those chicken coops on the tops of their car and drive across Texas don't do that anymore. They and their chicken coops can come right on my airplane." Breyer noted they certainly did, and he told PBS, "No one says it's fun, flying in an airplane filled with chicken coops. But nonetheless, if people want to pay the low prices for that kind of service, they should have the opportunity to do it. That's what had to happen in Texas, and now the object of the hearings was to ask why shouldn't that be true everywhere?"

Today, of course, it *has* become true everywhere, and the cabins of regional jets are jam-packed with chicken coops to prove this. I reached out to Justice Breyer to see if he has further thoughts on the effects of deregulation, but word came back that his views have not changed. In fact, most of those who fought for deregulation still support the principle, if not all the results.

Upon Kahn's death in December 2010, the *Wall Street Journal* ran this headline: "Stuck in an Airport? Blame Alfred Kahn (1917–2010)." Although he died just as I began researching this book, I contacted several people who worked closely with Kahn on deregulation: Michael Levine, Severin Borenstein, Diana Moss, and of course, Justice Breyer. They all maintain Kahn was an uncommon man, exceedingly bright and generous. But opinions vary slightly on how he viewed the aftereffects of deregulation.

Here are Kahn's own words, in testimony before Congress in 2000: "There are, I think, two things to be said about the fact that [a deregulated airline industry] has also been accompanied by a marked increase in discomfort and congestion. First, that it was precisely the failure of regulation to offer travelers a low-cost/lower-quality product that was its greatest failure. And second, that this deterioration in the quality of the air travel experience is a consequence, in important measure, of the failure of government to provide the optimal infrastructure—specifically, air traffic control and airport capacity—and to price it correctly."

The arguments persist on whether some form of regulation should be reimposed on the airlines. But one thing is certain: there is no going back entirely to 1978.

Brave New World of Competition

In August 1978, as that debate was raging in Congress, *Time* published a cover story—"New Era in the Air: Cheap Fares, Crowded Flights"—depicting a sardine can with wings. How crowded? Passenger load factors for U.S. carriers reached 61.5 percent systemwide that year. Consider that four empty seats out of every ten was considered "crowded" thirty-four years ago; in 2010 the domestic industry average had reached 82.1 percent.

Time opined: "Never has the future been less certain. The U.S. airline industry has been treated like a semi-monopolistic public utility, with routes and fares controlled by the Civil Aeronautics Board, which has sought to avoid over-competition and ruinous price wars. Now, President Carter seeks to free the airlines from Government economic controls entirely and allow them to fly anywhere at any time and charge any price, no matter how ridiculously low." The article itself is a bit of a time capsule, with its references to "stewardesses," call center wait times of "up to ten minutes," and hot versus cold in-flight meals. Virtually no one accurately predicted the explosive passenger demand of the deregulated industry, with *Time* stating: "In the future, passenger growth will be somewhere between 6 percent, which is the historic average, and 10 percent." In fact, in 1978 the U.S. airline industry carried 275 million passengers; that number would increase to 720 million by 2010, a rise of 162 percent.

What's particularly striking about the birth of airline deregulation was how it was midwived by the left, not the right as many assume today. Democrats controlled the White House and both houses of Congress in 1978, and the strongest proponents for unleashing free-market forces were President Carter and Senator Kennedy. And among those who testified on its behalf was Ralph Nader. Thirty-three years later, I asked Nader if he has any regrets, and he said, "In terms of competition, we supported it." But he stressed there were two underlying assumptions: the first was the implementation of stringent safety oversight, and the second was the implementation of stringent antitrust laws. Unfortunately, Nader maintained, both promises have been broken.

I visited my fellow FAAC member the economist Severin Borenstein, who began working for Kahn at an early age. We sat in Borenstein's office at the University of California at Berkeley, and he reminisced about the man who arguably had the greatest impact on the U.S. airline industry since the Wright brothers. Knowing that he remained in touch with his mentor Kahn until his death, I pressed him on what the architect of airline deregulation really thought of the end product more than thirty years later. "Fred said it didn't work out the way he thought it would," Borenstein acknowledged. What could not be foreseen was how legacy airlines would use tools such as frequent flyer programs, travel agency override bonus commissions, and dominance of hub airports to thwart competition: "Deregulation did not mean suspension of antitrust laws. Fred was quite vocal about this. The [Department of Justice] is where you sort of expect more oversight."

In two key areas—policing predatory pricing and overseeing mergers—Borenstein said the Justice Department has been "a real disappointment." Even so, he said Kahn did not recant: "He was still a believer in deregulation. It was a big improvement over regulation. But most of the fare decline has been due to higher load factors."

Today all domestic airlines and all domestic airline executives are strong proponents of deregulation, though that wasn't the case in 1978. Airlines for America, the primary trade organization for U.S. carriers, cites the dramatic growth of the industry and the plethora of low fares as proof positive of deregulation's success. "Even the carriers that have lost money still prefer to compete, because competition drives the market," says John Heimlich, chief economist for the A4A. "Reregulating means limiting competition and compensating airlines on cost."

Lost amid the deregulation debates are underserved regions of the country, particularly since, as the nonpartisan public policy organization Demos reported, more than one hundred communities had lost air service during the past decade, a condition the federal government has sought to address through the Essential Air Service program, which funds airlines.[1] In 2010, Severin Borenstein and I were asked by the FAAC to examine its usefulness, and facts quickly emerged: there is waste in the system, yet some communities, particularly in rural regions, remain dependent on EAS. One year later, EAS would be at the center of the partisan congressional feuding that cut off funding for FAA reauthorization in the summer of 2011.

The Arguments for . . .

Michael Levine is that rare individual who can teach—and can do. A respected law professor at Yale and New York University, he worked side by side with Kahn at the CAB and subsequently worked in the industry, as the CEO of New York Air and as executive vice president at Northwest. He lives not far from me in Connecticut, and we got together to discuss all things airline for hours on end. During our chat I mentioned that I've never owned a car that was not an American make, and when we walked out to his driveway to inspect my Pontiac Vibe, he noted it was a joint product of General Motors and Toyota.

Levine also pointed out that in the 1970s nearly all airlines fought deregulation, yet once it began "they decided anything goes." But he stressed that the original intent was not to restore the airlines to a laissez-faire world reminiscent of nineteenth-century England, but to ensure they behave like other businesses. And other businesses operate under a mixture of free markets and government regulation. Levine stated that the specific aims of airline deregulation were to do with "airline-specific regulation" damaging to consumers, such as control of routes and pricing; however, other standard retail sales obligations—such as disclosure of fares—would remain in effect.[2]

Yet many airline executives clearly feel that *any* form of government intervention is obtrusive; as former Continental Airlines CEO Gordon Bethune says, "It's the most regulated deregulated industry in the world." Passenger advocate Charlie Leocha agrees to an extent, by maintaining the airlines remain a "semiregulated" business, similar to electric utilities, because the government provides air traffic control through the FAA and oversees airline certificates through the DOT. "It's definitely a public-private enterprise," he says.

Hubert Horan has written extensively on the subject, and prior to one of our conversations he wrote extensively to me about it as well: "I fundamentally believe that deregulation was a hugely positive beneficial thing for the industry and consumers, although I can point out lots of minor implementation flaws and errors." He added: "Deregulation doesn't stop companies from doing dumb things, and doesn't stop fuel prices from increasing. The benefit of deregulation is that it limits the ability of dumb—or politically powerful—companies from running to the government for protections from the consequence of dumb decisions."

. . . and Against

Kevin Mitchell, the president of the Business Travel Coalition, asks this question: "Under deregulation, how many stakeholders are happy? Investors? Management? Labor? Customers? None of the above." Consumer advocate Kate Hanni goes even further: "I think it has ruined the airline industry. It's nearly impossible for anyone to make money for more than three quarters. In general, the stockholders and low-level workers and passengers have borne all the risk, but not the executives."

In 2009 I was contacted by Demos for a research project titled "Flying Blind: Airline Deregulation Considered." The report concluded that the airline industry is more concentrated than ever, and "most of the major U.S. airlines" are relying on an increase in outsourcing, service cutbacks, hidden charges, employee wage and benefit reductions, and consolidation, with "the hope of surviving long enough to be in a position to turn a profit and expand again during a future economic recovery."

Not surprisingly, many labor officials are unhappy with how deregulation has played out for the airlines, since by some estimates the industry worldwide has lost nearly one million jobs over the last thirty years. Robert Roach of the International Association of Machinists says it has spurred many job losses, and therefore customer service has deteriorated as well.[3] He adds, "It's been consistently downhill since 1978. With airlines it's always, 'We have to compete with the lowest cost structures.'" For his part, James P. Hoffa of the Teamsters says deregulation has led to a continual decrease in standards, which is stressing aviation.

Ironically, one of the loudest voices belongs to the man who arguably was more successful than any other airline executive at parlaying deregulation into corporate success: Robert Crandall. Under his stewardship, American Airlines didn't just have a growth plan, it had a Growth Plan, articulated and shared with the world. While deregulation helped kill former giants such as Pan Am, TWA, and Eastern, it also helped propel midrange carriers into today's industry titans: United, Delta, and certainly American. Yet Crandall has come full circle: just as in the 1970s, today he is opposed to deregulation and is probably the industry's strongest proponent for reregulating.

He told me, "Everything really was relatively fine up until 1978. Now I do think the CAB was unduly restrictive. I think we failed to appropriately regulate labor. So I think the CAB should have permitted more competition, and I think labor should have been regulated. But the whole notion of a regulatory structure that says the fares should be about the same on a per-mile basis . . . and that efficient operators should be able to earn a return on capital is appropriate for a utility-equivalent business."

Other voices are drowning him out, notes Professor Paul Dempsey, an expert on aviation law who refers to Crandall as "the brightest man I ever met in the airline industry." He attended a Wings Club luncheon in New York City in 2008, when Crandall spoke about the industry's need for "a dollop of regulation, along with new government policies and appropriate investment." Several years later, Dempsey said it was extraordinary the remarks have attracted so little attention, as if they were never uttered.

The R-Word: Reregulation

Let's note an important point: it's not that airline executives are opposed to government intrusion; they're just opposed to it when it doesn't help them. In my travel column for USAToday.com in 2008, I addressed this phenomenon and noted that over the years airline executives have always been eager to welcome government intrusion when it works in their favor, a phenomenon best described as *hypocrisy.*

Over the course of researching this book I discussed with dozens of industry experts what some form of reregulation might look like. A time trip to 1978? Or minor tweaking, such as Crandall's "dollop"? But a problem emerges. With many airline executives and analysts, discussing the free market is akin to discussing Torah or the Koran or the New Testament, and heresy is not tolerated.

Dempsey has noticed it as well: "A lot of policy is driven by ideology. It's almost religious. . . . Communists were viewed as having a nearly religious passion—but free market economists have the same passion. Both are equally unwilling to look at the failures of their systems." Referring to deregulation as a "mantra," Dempsey says, "It has to do with belief rather than rational thought." Even a strong free-market proponent such as Borenstein agrees.

Not all those who recommend some form of reregulation necessarily suggest the U.S. government get back in the business of determining fares. Dempsey notes this is rife with challenges: "Do you set the fares so that you make United unprofitable and make JetBlue profitable? Or so that you make United profitable and JetBlue obscenely profitable? This is politically risky." He adds, "Airlines are doomed to fail in the long term."

Others offer interesting perspectives as well:

- James Lardner of Demos: "Prior to 1978, under Alfred Kahn, there was a lot of dynamic stuff happening. Which suggests there is a middle ground between regulation and complete deregulation. . . . The airlines tend to get a disproportionate power through reservations systems, frequent flyer programs, fortress hubs, etc. So you can treat them like a public utility, either wholly or on routes where there is no competition."

- Labor leader Pat Friend: "Complete reregulation is never going to happen. I think we need open-minded people to sit down and say, 'Where can we tweak this?' The airlines are supposed to serve the public and serve the economy and not be allowed to run amok."
- Labor leader Robert Roach: "We need a mild form of reregulation." This would include restricting aircraft maintenance to the United States, banning predatory pricing, and tightening workforce rules.
- Consumer advocate Kate Hanni: "I don't know if it includes rates and routes, or simply a comprehensive reregulation without rates and routes."
- Aviation law professor Paul Dempsey: "The mantra of the right is that the government can do no right and the market can do no wrong. . . . Still there is no meaningful discussion about reregulation. Things have to get a lot worse before they get better. So I guess things are not bad enough now. I guess it hasn't gotten bad enough yet if the secretary of transportation says reregulation is not on the table. I think deregulation has been a catastrophic failure. And I recognize that I'm in the minority. But I think I'm right. I wish I was wrong."

For insight I turned to the best boss I ever had in the airline industry, Harris Herman, who was president of the Pan Am Shuttle in a kinder and gentler time. He noted that be began his airline career fourteen years before deregulation, when it was a very friendly, low-pressure environment. He also believes the industry could not have stayed regulated until today (even though undoubtedly Pan Am would still be around).

Michael Levine believes any talk of reregulation falls prey to the Nirvana Fallacy, in which some believe flawed markets can be corrected by perfect regulation, while others believe perfect markets are marred by flawed regulation. The truth, he asserts, is that in reality we have only flawed markets and flawed regulation, and zealots on both sides place too much belief in the inherent powers of both markets and regulation.

I pointed out to him what I've uncovered about airline maintenance outsourcing and the FAA's failure to provide oversight. And Levine responded, "So your problem is you have found flawed markets and flawed regulation. And I'm not telling you that you haven't found it. But I'm telling you that when you write about it, you need to ask, is it really going to be any better with more regulation? Or is [New York senator] Chuck Schumer just going to be able to stand up and have a press conference?"

Point taken. Ultimately we may need to decide which entity we trust *less*—the U.S. government or Corporate America.

The Best Airline Investment Advice: Don't!

The reregulation argument is bolstered by the industry's dismal financial performance since 1978. Airlines make for rotten long-term investments, and with the singular exception of Southwest, U.S. airlines are particularly rotten over time. In fact, some argue that all forms of transportation have duped investors. Sir Richard Branson, the founder of Virgin Atlantic Airways and its sister carriers located around the world, as well as minority owner of Virgin America, put it best. When asked how to become a millionaire, the man who had earlier founded Virgin Records replied: "There's really nothing to it. Start as a billionaire and then buy an airline."

An awful lot has been written about the economic state of the industry, and analysts crank out more material every week. It's impossible to address virtually any aspect of the business—passenger service, labor, maintenance, safety, security—without providing an economic perspective. All roads lead back to airline costs, and what airlines are doing to cut those costs.

Throughout its history, the industry has weathered economic peaks and valleys, but since deregulation it's positively been a roller-coaster ride. For domestic airlines, there were losses of $23.7 billion in 2008 and $2.5 billion in 2009. Then a surprise return to profitability in 2010 due to increased demand and higher fares, followed by more losses for most U.S. carriers in early 2011. Along the way, fortunes have been made—and unmade. Travel pricing expert George Hobica points out, "All those many billions of dollars have been lost. Where did they go? . . . Who really has lost?" He counts employees, investors, and banks among the victims, and says, "It's incremental. ATM fees probably went up half a cent because banks took a bath with airlines."

Many aviation pundits believe that any discussion of airline stocks begins and ends with Warren Buffett, arguably the world's shrewdest buyer and seller of capital. In a letter to shareholders of Berkshire Hathaway in February 2008, he famously stated: "The worst sort of business is one that grows rapidly, requires significant capital to engender the growth, and then earns little or no money. Think airlines. Here a *durable* competitive advantage has proven elusive ever since the days of the Wright Brothers. Indeed, if a farsighted capitalist had been present at Kitty Hawk, he would have done his successors a huge favor by shooting Orville down."

Buffett went on to detail his "shame" in buying preferred stock in USAir (later US Airways) back in 1989, just before "the company went into a tailspin." He further noted, "The airline industry's demand for capital ever since that first flight has been insatiable. Investors have poured money into a bottomless pit, attracted by growth when they should have been repelled by it." But Buffett won't make the same mistake twice: "I have an 800 number now which I call if I ever get an urge to buy an airline stock. I say, 'My name is Warren, I'm an air-aholic,' and then they talk me down."

Former American Airlines CEO Bob Crandall maintains investors and airlines are simply a bad match: "No, you can't make money in the airlines. You can trade. But from an investment point of view, they're a catastrophe. They always have been a catastrophe." He's echoed by legions of analysts, journalists, and everyday investors. As columnist Joe Brancatelli noted on Portfolio.com: "Say whatever else you will about airlines, but one fact is incontrovertible: They make lousy investments. For about 20 years, give or take a quarter or two, only the shorts have profited from airline stocks."

And with credit ratings, Airlines for America points out that *no* passenger airline—even the most efficiently run—enjoys a rating of A- or better:

Southwest	BBB-
Alaska	BB-
Allegiant	BB-
Delta	B
United	B
American	B-
JetBlue	B-
US Airways	B-

As for these titanic losses, economist Severin Borenstein says, "The legacy airlines do not have a workable business plan. And they compensate by using their power to fence out competitors and continue to sustain high costs. In some cases they don't have the management capability and in other cases they don't have labor union cooperation." He adds, "We are seeing a trend of legacy airlines entrenching long-run."

But there is one notable exception. In an industry known for burning through barrels of fuel *and* barrels of red ink, Southwest Airlines has consistently, reliably, and repeatedly done what *no* other U.S. air carrier has managed to do since Richard Nixon was president: make money. At the end of 2010, the Dallas-based company marked its thirty-eighth consecutive year of profitability.

Others have made a comfortable living betting against airline profitability. As one major airline CEO points out, "There's a dirty little secret about industries such as this one. There are always people who profit and benefit from dysfunction."

Grounded Forever: Airline Bankruptcies and Shutdowns

In late 1991 I left the airline industry for good after the Pan Am Shuttle was sold to Delta and Pan Am itself succumbed; that same year saw the shutdowns of two other major domestic carriers, Eastern and Midway. That fall I applied for unemployment insurance and traveled to the New York State Department of Labor office in Flushing, Queens, close to both LaGuardia and JFK airports. I soon found that in order to expedite our claims, there were two lines in the office; one handwritten sign read PAN AM & EASTERN and the other sign read ALL OTHERS.

Since deregulation, bankruptcies have become ubiquitous in the airline business. According to Airlines for America, there have been 189 bankruptcy filings just since the deregulated era began in 1979. The A4A points out that prior to 1978, bankruptcies were "extremely rare," with the CAB arranging "marriages" between failing carriers such as Northeast and surviving carriers such as Delta.

Over the last two decades, most major domestic airlines have passed through Chapter 11 reorganization, including America West, Continental, Delta, Northwest, TWA, US Airways, United, and—as of last year—American. Some majors have filed more than once; US Airways, for example, sought Chapter 11 protection in 2002 and again in 2004. Amazingly—or perhaps not so amazingly—both Northwest and Delta filed on the exact same day, September 14, 2005, less than three years before those two carriers merged.

Among the largest carriers, the notable exception has been the Dallas-based airline Southwest, which has never filed. Last year former Southwest and Braniff CEO Howard Putnam said, "Now almost every airline has been through it. All but American, and they probably wish they had." He could be right: back in 2003, American CEO Don Carty was forced to abruptly resign after a Securities and Exchange Commission filing revealed he had received a $1.6 million bonus. The *New York Times* reported that American had made a $41 million pretax payment into a trust fund created to protect the pensions of 45 executives.[4] All this occurred as Carty was negotiating with American's unions for $1.62 billion in concessions.

In the years after the terrorist attacks in 2001, another wave of bankruptcies swept the airlines. But as industry expert Hubert Horan points out: "All the post-9/11 bankruptcies were due to the reckless overexpansion of the late 1990s, even though the airlines falsely blamed their losses on some combination of Osama bin Laden and evil unions." In recent years U.S. airlines have been disciplined about "restraining capacity" by parking airplanes in the desert, laying off employees, and reducing the number of available seats. Airlines for America reported that in 2009 domestic seating capacity had fallen to its lowest point since 1942, when the airline industry was mobilized by the War Department.

The Same Old Victims: Employees and Passengers

Kenneth Goodpaster, an academic who specializes in business ethics, wrote a case study for Harvard Business School titled "Braniff International: The Ethics of Bankruptcy." I asked Goodpaster if employees should be rewarded when and if the company does turn itself around; in other words, shouldn't executives give back the givebacks? "It certainly would be a nice gesture," he said. "It could be seen as appreciating that labor gave back. As long as the company's financial health could sustain such increases. Is it a legal obligation? No, of course not. Is it a moral obligation? I'm not even sure. But it could be the morally admirable thing to do."

Givebacks may be too much to seek. For workforces that have seen their jobs dissolve and their paychecks reduced, the salt in the wound has been that not everyone shares the pain, particularly in airline executive suites. In 2007 a lengthy article analyzing bankruptcy and executive compensation in the *Northwestern University Law Review* found that "recent reform efforts to limit executive pay in bankruptcy are largely misplaced." However, even this report added that data show "disproportionately large grants" to CEOs in the years leading up to Chapter 11 filings; the report noted that such grants are better explained as severance payments, and courts should give them further scrutiny.

That same year, airline labor unions participated in the "Enough Is Enough" rally on the Mall in Washington, D.C., to protest out-of-control management self-interest. The Aircraft Mechanics Fraternal Association stated: "United Airlines' board bestowed $39.7 million in compensation on CEO Glenn Tilton in 2006, despite the airline's continued poor performance. AMFA recently used the UAL shares the union owns to give the UAL board a no-confidence vote." Columnist Joe Brancatelli is blunt in his assessment: "This was a guy who knew zippo about the airline business—and after 120 days he bought into the conventional wisdom and he destroyed the employees' equity."

Many hard-core free marketers scream that bankruptcy laws protect the weak from the Darwinian effects of capitalism. But that hasn't stopped airline CEOs from profiting at others' expense. "The point of bankruptcy laws is to get the most for the creditors who have been fucked and to keep the company alive in the interim," says Hubert Horan. He also faults preferred creditors, such as Boeing and Airbus and jet engine manufacturers, for their role in airline bankruptcies: "It's like giving Tony Soprano control of all the waste disposal in North Jersey." He is especially harsh in discussing United's reorganization, which lasted from 2002 to 2006 and was particularly contentious. Tilton came under heavy fire because despite widespread layoffs and the cancellation of United's pension plan, Forbes still ranked him as the highest-paid U.S. airline executive in 2005. "It was a group of clearly conflicted interests," says Horan, who claims the court effectively "transferred assets to the personal account of Tilton." Meanwhile, Horan contends the result was "a glut of capacity" for years that harmed the entire industry.

Bernadette McCulloch of the Teamsters speaks for many labor representatives when she argues that Chapter 11 has become a green light for airlines to wipe clean all obligations to employees: "The minute an airline goes into bankruptcy, that's when they really start outsourcing. The judge can nullify your contracts so they basically hold you hostage." McCulloch notes that a long, drawn-out battle was waged to preserve a handful of employees at Frontier Airlines, but saving those 121 jobs came at a great cost. She also notes that railroad employees keep their pensions for life, even if they move from one company to another and even when the industry consolidates, a policy that airline executives would never allow.

Of course, the other victims in airline bankruptcies are passengers, and sometimes even whole communities. A decade later, St. Louis still hasn't recovered from the loss of TWA: the bankrupt carrier's former hub in St. Louis saw a reduction in total passenger traffic from 23 million in 2002 to 12 million in 2010. As passenger advocate Kate Hanni notes, many passengers don't realize their tickets become worthless when the airline goes bankrupt. Unfortunately, rival airlines often do not honor tickets from an airline that is shutting down, and members of frequent flyer programs can be left with worthless mileage as well. Oh, and don't count on travel insurance. Many underwriters maintain "black lists" of airlines they consider financial risks, so carriers undergoing Chapter 11 bankruptcy reorganization are always verboten as risks on travel policies.

Consolidation: Enough Shrinkage Is Never Enough

Within a three-week period in 2010, I appeared before the Senate and then the House to testify at hearings examining the proposed merger of United Airlines and Continental Airlines. On behalf of Consumers Union, I strongly urged Congress to support passengers by weighing in against that marriage, on the grounds there would be less choice and fewer flight frequencies for passengers, as well as a complete loss of service on some routes. As for airfares, it's an economic given that consolidation leads to higher prices; it's one of the few topics that virtually all economists agree on. And with fewer competitors, it's harder for any airline to resist matching an increase in fares or fees. But there are other ill effects as well, including reductions in service quality. After all, if there is no meaningful competition, what incentive is there to improve service or launch new service initiatives? Over time, consolidation also has a chilling effect on the launch of new start-up airlines, and consumers lose out on the benefits of additional service and lower fares there as well. Ultimately the threat of widespread disruptions increases, because the loss of a single carrier—whether it's due to bankruptcy, a labor action, or an FAA shutdown—could cripple large sections of the nation. Eventually a "too big to fail" scenario arises. And finally, each approved merger and acquisition leads to more rubber-stamping of additional couplings, as airlines argue that they should be afforded the same privileges given the last merger partners.

I provided detailed analysis of how fares rose and service decreased in case after case when the airline mergers were approved. Of course, Congress did nothing and the Justice Department approved United/Continental that summer, just as it has approved virtually every airline merger put before it in recent memory. That fall, I entered an FAAC Competition Subcommittee meeting in Denver chaired by Glenn Tilton, CEO of United, who was speaking enthusiastically about post-merger plans. I feigned naïveté and asked if the deal had been approved, to which Tilton replied, "Yeah, we'd like to thank you for your help, Bill."

First, some needed context. The airline consolidation trend is nothing new; in fact, it's as old as the industry itself. Ever wonder why it's *United* Airlines? The company was formed by melding a series of smaller airmail and passenger carriers. Similarly, Delta didn't take shape until its acquisition of Chicago and Southern Air Lines in 1953. And back in 1962 *Time* reported that Juan Trippe of Pan Am was seeking a merger with Howard Hughes's Trans World Airlines so that "Pan Am World Airlines" would become the "chosen instrument," a single U.S. carrier designated by Washington to operate overseas.

Within a brief period between 1985 and 1987, the U.S. airline industry saw a wave of fourteen domestic mergers, including the acquisitions of venerable players such as Eastern, Frontier, Ozark, Piedmont, Republic, and Western. (Such a list can be confusing; commercial aviation has a long history of recycling brand names such as Allegheny, Frontier, Midway, Piedmont, National, Republic, and even Pan Am. For years I told industry veterans, "I used to work for National Airlines. No, not that one—the other one.")

Now consider what has occurred in just the last decade:

- 2001: American acquired TWA's assets through bankruptcy
- 2005: US Airways and America West merged
- 2008: Delta and Northwest merged
- 2009: Republic acquired Midwest and Frontier
- 2010: United and Continental merged
- 2011: Southwest and AirTran merged

Put another way: in 2004, aviation journalist Jerome Greer Chandler noted that of the roughly forty domestic U.S. airlines that existed in 1929, just two—American and Northwest—continue flying under their own names. By 2010, Northwest was absorbed into Delta, leaving American as the lone survivor from that era.

The most honest assessments on consolidation usually come from the financial community. The website InvestingDaily.com stated in 2010: "Investors love airline mergers because consolidation means less competition and less competition means fuller planes (aka higher 'passenger load factors'), higher airfares, and more profits for the remaining players." Some, such as Dan McKenzie of Rodman & Renshaw, believe consolidation has made for a healthier industry. But nearly all economists agree it also leads to higher fares. "It's inevitable," says airfare expert Bob Harrell. "You've got fewer people making the decisions. You go from ten to eight to six to four. They're going to behave in an oligopolistic way. The flip side is they have to do something to get a return to profitability."

And there's more consolidation to come. Analyst Helane Becker of Dahlman Rose & Company predicts that there will be further shrinking within the competitive regional airline field. However, she defies conventional wisdom in that she does not predict an American/US Airways merger, and instead sees a Big Three consisting of United/Continental, Delta, and an ever-expanding Southwest.

And make no mistake: airline CEOs love the thrill of the deal. There are analysts who believe we're facing a shrinking airline industry because competing execs love comparing the size of their expanding . . . egos. Consider that within just two years, the title of "Largest Airline in America" was swiftly passed from American to Delta/Northwest to United/Continental.

Back in the mid-1990s I asked the head of a major domestic carrier if codesharing marketing agreements wasn't providing all the benefits of mergers, but without the messiness. This practice allows two or more airlines to sell seats on each other's flights, using multiple flight numbers. Heck, passive DOT regulators had even begun providing papal blessings in the form of antitrust immunity; that way "competing" airline executives could do what they had always done—illegally discuss price fixing—only now do it legally. What's not to like about codesharing if you're the CEO? The airline executive summed it up for me off the record: "You can have pure sex or you can jack off. That's kind of jerking off, okay? I mean, it really is." I've never gotten a more honest answer from the head of an airline.

With one exception. Former American Airlines CEO Crandall has seen an awful lot, both before and after deregulation, and here's his take: "Up until 1978 we limited the number of participants by regulating. In the last three or four years we've limited the number of participants by approving every goddamn merger that has happened. So we're making a social choice, which I have long felt is the wrong social choice, and that is we're going to let the industry regulate itself by consolidating to the point where there are so few competitors that the price structure will support the costs. And that's the whole story of the airline business."

Goodbye to Low Fares?

A year after we first met in Congress, I talked consolidation with Tilton. "There's not enough room left," he told me. "What I think will happen is we'll have two fundamentally different business models. There will be network airlines, and they will be challenged on the periphery by Spirit, Allegiant, and Virgin America."

However, if Tilton is right and Southwest/AirTran and JetBlue morph into quasi-network carriers, that means there will be less of a low-cost airline presence in the United States. And this will represent a huge step backward for consumers. That's troubling from a consumer perspective; Diana Moss, director of the American Antitrust Institute, puts it like this: "It's hard to say no to subsequent mergers when you've already said yes to the big guys. . . . When somebody decides to buy JetBlue or Frontier, they'll allow it. And that's unfortunate—because we'll wind up with three airlines, and that's a scary prospect."

Unlike many organizations that focused primarily on consolidation of legacy carriers, the AAI raised concerns about the "novel issues" generated by the first major merger of low-cost carriers, Southwest and AirTran. As Moss says, "The real worry is if you take out the low-cost carriers—then where are we? The prices creep back up to the legacy levels. As there is more consolidation it's harder to enter the airline industry. Who wants to compete against eight-hundred-pound gorillas?"

During those United/Continental hearings, then congressman James Oberstar—a Democrat from Minnesota—fought against consolidation, but within months he was voted out of office after thirty-six years (and beaten by a former Northwest Airlines pilot).

As for the future wedded bliss of United and Continental, former Continental CEO Gordon Bethune is painfully blunt in recalling an earlier occasion when United attempted to acquire his old carrier: "I told them, you can't run water." He believes the real challenge now is cultural, as the two company workgroups are slugging it out to determine which group is stronger. But he pulls no punches in detailing United's shortcomings: "It's the same malaise as [at Continental] in 1994 and 1995 when I got there. Employees treated each other like shit. We turned it around. They don't have that culture at United." He should know: before Glenn Tilton became CEO of United, the company's board approached Bethune about taking the job.

What's particularly compelling is that the same pattern is repeated over and over and over: Airline CEOs promise Congress, local politicians, unions, and the media that they will not downsize in a given city and they will not cut jobs. And then once the merger is approved, they do just that. It was demonstrated yet again last year, when Minnesota politicians howled after Delta announced it would move several hundred training and technical jobs from Northwest's former home state to Delta's hometown of Atlanta. Of course, there's a bit of a dog-bites-man aspect to such stories. Any analyst could have pointed out that Delta did not need two sets of pilot training centers, flight attendant training centers, and flight simulators. Consolidation is all about economies of scale, so why the shock when Delta did just that—consolidate?

The Politics of Merging

As for the "too big to fail" argument raised by me and others, some experts believe we're rapidly approaching that threshold, while others think we're already there. I agree with this second group, and think that if United/Continental is in danger of collapsing, then the government will have to step in and prevent that. Unfortunately, there is a lot of misinformation supplied by the media when airlines consolidate or shut down, as detailed in chapter 12. But Horan points out the "experts" who are quoted are often in the tank as well: "The Wall Street analysts never saw a merger they didn't like because they want the [merger-and-acquisition] fees."

Loyalty program expert Tim Winship points out that "consolidation is all about pricing power, and pricing power, by definition, benefits the airline, not the consumer. So all of this chatter about benefits to consumers is overstated at best and downright false at worst." He continues: "What additional value does a consumer get after the merger? For every benefit there's probably a countervailing negative. Ultimately it takes a competitor out of the market, and anyone who says that won't lead to higher prices has never studied economics. The net effect is negative."

In June 2011 the American Customer Satisfaction Index stated that airline mergers typically have a "destructive effect" on customer satisfaction, citing the examples of US Airways and Delta following their respective mergers. Although the jury remains out on United/Continental and Southwest/AirTran, the historical record is not good from a passenger perspective.[5]

In early 2007 US Airways warned of "hiccups" during the integration of the computer reservations systems used by US Airways and America West during their merger. In fact, during the melding of these systems, only about half of the new carrier's flights arrived on time. We've come to expect such problems when airlines merge, but employees point out there are potential safety risks as well.

Philomena Larsen was a longtime employee of Northwest Airlines, but she began her career at Republic Airlines in the mid-1980s just prior to that carrier's acquisition by Northwest in 1986. She calls that merger "horrendous" and notes there were widespread problems with lost bags when the two airlines' baggage tracking systems were melded. But Larsen also points out that mergers and acquisitions have a safety component as well: employees unfamiliar with new aircraft, ground vehicles, ramp equipment, and baggage belts are often rushed through training. In the case of Northwest/Republic, Larsen says the integration of two very different weight-and-balance systems meant employees were learning the new system on the job—a dangerous proposition for such a critical safety function. "We were lucky there wasn't an airplane accident," Larsen recalls.

3

Collusion and Confusion: How Airlines Don't Play by the Rules—and How Passengers Pay

ROBERT CRANDALL, CEO OF AMERICAN AIRLINES: Raise your goddamn fares 20 percent! I'll raise mine the next morning!

HOWARD PUTNAM, CEO OF BRANIFF AIRLINES: Robert! We—

CRANDALL: You'll make more money and I will too!

PUTNAM: We can't talk about pricing.

CRANDALL: Oh, bullshit, Howard! We can talk about any goddamn thing we want to talk about!

—*Taped telephone conversation submitted as evidence by U.S. Department of Justice in Antitrust Division civil lawsuit, 1993*

AIRLINE EXECUTIVES LIKE TO TALK about the need for a "level playing field." On the other hand, many of these same execs do their best at all times to unlevel the corporate playing field. And it can take many forms: industry lobbying (both on and off the record), predatory pricing and dirty tricks against competitors, biased distribution that taints airfare shopping, codesharing and airline alliances, and antitrust exemptions and violations.

The key is buying influence. The transfer of wealth—and we're talking "regressive distribution" here, where money flows from the poor to the rich and not vice versa—remains the most important and underdiscussed issue in American life, at least until the Occupy Wall Street movement shone a light on it. There is a growing library of documentation illustrating how government and corporations work hand in hand to redistribute wealth through "trickle up" economics, under both Republican and Democratic administrations and majorities in Congress. In February 2011, CNN.com compared Internal Revenue Service data from 1988 and 2008 and found that, adjusted for inflation, the average American income fell slightly from $33,400 to $33,000. But during that same twenty-year period, the incomes of the richest 1 percent of Americans rose by 33 percent.

These issues not only cross party lines but permeate the entire American political system, in all branches and at every level. In January 2010 the U.S. Supreme Court officially granted the right of free speech to corporations, by lifting all restrictions on how much companies can spend to support and assail political candidates. Meanwhile, executive compensation soars.

The flow of money to politicians from the aviation industry remains a tangible fact of life. I began joking about it during FAAC meetings: the airlines have a permanent lobbying presence in Washington; consumers take what they can get from passenger advocates who can't compete on staffing and budgets and contributions.

Consumer advocates point out the airlines have virtually unlimited resources, so those defending the rights of passengers are at a disadvantage What's more, it's been well documented that there's a revolving door between the airlines and government service. In fact, one report from Public Citizen cited dozens of former higher-ups at the Department of Transportation, the Federal Aviation Administration, the National Transportation Safety Board, White House, and Congress who shilled for the airlines—and some who made the journey in reverse.

Make no mistake. Airlines constitute one of the most powerful of all Washington lobbies, totaling nearly six hundred individuals. In fact, from 1998 through 2011, air transport ranked fifteenth in the nation in dollars spent in an attempt to buy influence, amounting to a total of $793 million.[1]

That's some army of lobbyists, and unfortunately they're often laboring against the best interests of customers and employees. First of all, there are the one hundred or so folks working for Airlines for America. Then there are independents such as Linda Hall Daschle, former Miss Kansas, who has lobbied for aircraft manufacturers Boeing and Lockheed-Martin, and just happened to have served as acting administrator of the FAA (customers first!). Her husband, of course, is former senator Tom Daschle of South Dakota, the Democrat who served as Senate majority leader. That $15 billion bailout for the airline industry, rushed through the Senate just eleven days after the 9/11 attacks? Yep, at the time Linda was an airline lobbyist and Tom headed up the Senate. Her other clients have included American Airlines, Northwest, and L-3 International, a company the FAA paid for airport security scanners that, well, *happened* to leak radiation. After President Obama tapped Tom for a cabinet position (Daschle later withdrew), Salon.com profiled the couple as "The Daschles: Feeding at the Beltway Trough." That article quoted *Rolling Stone* contributor Matt Taibbi: "In Washington there are whores and there are whores, and then there is Tom Daschle."

Obviously there are also countless lobbyists working for individual airlines. And representing outsourced maintenance we now have the Aeronautical Repair Station Association staking its claim in Washington as well. ARSA recently launched the "Positive Publicity Campaign Plan," which includes an allotment of $1 million annually for targeted public relations outreach and creating "a tactical plan to deliver those messages to key audiences" (including policy makers) over three years.

Hope and Change for Passengers?

During the second term of President George W. Bush I saw countless bumper stickers with the simple message *1-20-09.* Yet only one month into the Obama administration I spotted a sticker touting *1-20-13.* We seem to have entered an era of permanent teeter-totterism in American politics: the populace is perpetually angry with incumbents, and the next government-in-exile carps from the wings. Yet the most pervasive issue of our time—the wholesale purchase of democracy by corporate interests—is never discussed in a meaningful way by *either* party.

So does it matter which party is at the helm? I reached out to Ralph Nader for edification; after all, who better? He claimed it doesn't matter if Democrats or Republicans are in power: "It's a permanent government. The airlines and the manufacturers pick the head of the FAA. Have you ever noticed there is never a contentious hearing for the FAA administrator?"

During the 2008 campaign, the Teamsters were early and strong supporters of Senator Barack Obama rather than Senator Hillary Clinton or any of the Republican contenders, perhaps not surprising given both the Clinton and Bush records on free trade agreements such as NAFTA. In a letter addressed to the Teamster Aviation Mechanics Coalition, the future president stated: "The practice of outsourcing aircraft maintenance overseas raises security concerns and pits our skilled mechanics making a middle class living against less skilled, less well protected workers abroad. I applaud your efforts to organize a strong union at United Airlines, and look forward to working with you on the critical issue of outsourcing now and in the years ahead."

Unfortunately, the outsourcing crisis has only gotten worse.

As labor leader Tom Brantley says, "Senator Obama supported oversight of repair stations. President Obama doesn't." And let it be noted one year after inauguration, the forty-fourth president's secretary of transportation tapped the chairman of United Airlines to head the FAAC Competition Subcommittee.

Some argue Obama has been an improvement, however, and point to DOT secretary LaHood's passenger rights regulations as proof. "There's been a decent amount of change," says Paul Hudson of the Aviation Consumer Action Project. He asserts Republican views toward passengers were unduly harsh: "The DOT secretary would tell Congress that market forces would address all problems. Or else people don't have to fly."

There's also no denying that some presidential administrations have been more proactive than others on airline safety. For example, an analysis conducted in 2010 by News 21–Center for Public Integrity found the NTSB "issued significantly fewer recommendations for improvements" under President George W. Bush. Here's the tally of average annual NTSB recommendations for improvement under five consecutive White House staffs:

Carter	384
Reagan	445
G. H. W. Bush	417
Clinton	329
G. W. Bush	155

In response to how both Republicans and Democrats have served the interests of the airline industry rather than the interests of passengers, there is strong evidence such influence may affect even the most high-profile politicians. Industry veterans recall that Senator John McCain was an outspoken advocate for passenger rights reform in 1999 and sponsored the Airline Passenger Fairness Act; I can recall a telephone conference with him on the topic. Then, in June 1999, he suddenly switched course and supported a watered-down voluntary pledge from the airlines, and this *Washington Post* headline sums it up succinctly: "A McCain Crusade Faded as Airlines Donated." Yep: that month the airline industry proffered $226,000 in "soft money," including $85,000 from American Airlines for the National Republican Senatorial Committee. The *Post* also reported that Phoenix-based America West (now Phoenix-based US Airways) was "one of McCain's top benefactors." Hudson recalls being stunned by McCain's actions: "He ditched the bill and said the airlines would address it themselves. Coincidentally he had gotten lots of donations from airlines. We've seen this movie before."

Allegations of unlevel playing fields apply not only to airlines but also to commercial aircraft manufacturers. In March 2011 the World Trade Organization issued a lengthy "dispute settlement" in response to a complaint filed by Europe-based Airbus, finding that Boeing had been unfairly subsidized by a host of U.S. government entities, including NASA, the Department of Defense, and the states of Washington, Kansas, and Illinois. However, one year earlier the WTO had found that the European Communities and the nations of Germany, France, the United Kingdom, and Spain had unfairly subsidized Airbus. Clearly when it comes to corporations bending the rules, there's plenty of blame to be spread on a global basis.

So how did a government of, by, and for the people suddenly become a government of, by, and for the S&P 500? "It's all caught up in this free trade/free market mantra," says airline academic Paul Dempsey. "You have high school graduates, blue-collar workers with no jobs. But that doesn't matter. They're told they can go flip burgers. What matters is that consumers are given lower prices." Dempsey believes Oscar Wilde's assessment of cynics also applies to economists: "They know the price of everything and the value of nothing."

Predatory Behavior = Higher Fares

An industry executive once put it best: "The hub airlines are like badgers in their lairs. They won't come out looking for you, but you're in trouble if you go in and screw with them." Many low-fare airlines have found this out—in some cases, too late. When a smaller airline starts flying on a major's bread-and-butter route, the result can be all-out war—and passengers are the casualties, with less service and higher fares in the end. The weapons of choice can include predatory pricing—temporarily offering rock-bottom fares to drive out low-cost competitors—or the majors can even engage in dirty tricks, such as poaching passengers.

For those who thought the legacy airlines had moved beyond such behavior, in March 2011 the Business Travel Coalition pointed out it was déjà vu all over again, this time with Delta's retaliation against low-cost Frontier Airlines for starting scheduled flights out of Delta's hub at Minneapolis–St. Paul International Airport.[2] Bob Harrell notes about predatory pricing: "You can smell a rat, but it's hard to prove."

Both the departments of Transportation and Justice have attempted to do just that. For context, it helps to review what the airlines are capable of, by illustrating how they have behaved until now. What's more, the myriad ways in which major airlines have attempted—both legally and illegally—to bankrupt low-fare airlines further hurt consumers by dampening enthusiasm within the financial community to support start-up carriers.

In 1993 an influx of predatory pricing charges were leveled by new-entrant airlines against hub-and-spoke majors. That July, Houston-based UltrAir shut down, claiming "illegal, anticompetitive, monopolistic, and predatory behavior" by the other hometown carrier, Continental; similar charges were leveled against Continental at Newark, New Jersey, by KIWI. Soon after, the Justice Department investigated claims that Salt Lake City–based Morris Air was being penalized by Delta's override policies, which rewarded travel agents for steering customers away from Morris. And the transportation secretary personally intervened when Reno Air cried foul against Northwest. The obvious irony is that all four of those start-ups soon went through mergers or bankruptcies (then again, so did Continental and Northwest).

An industry white paper by Clinton V. Oster Jr. and John S. Strong, "Predatory Practices in the Airline Industry," illuminates the Reno Air–Northwest brawl. The new guys began service between Reno and Minneapolis–St. Paul in 1992. Northwest had abandoned the same route one year earlier, but after Reno Air met with success Northwest announced new service. Pulling out all stops, Northwest also said it would match the low-cost carrier's fares; it offered bonus frequent flyer mileage, and those overrides for bookings out of Reno. Then Northwest lowered its *nonstop* fares between Minneapolis and the West Coast, an obvious effort to attract Minnesota-to-California-bound customers flying Reno Air's *connecting* service through Reno.

So far it seemed like good old-fashioned bare-knuckle capitalism. To the surprise of no one, the big guy won, and Reno Air discontinued its flights in June 1993 (by 1998, it was acquired by American). But that's where "free marketism" crosses over into "illegalism." Once Reno Air pulled up stakes, Northwest's fares on those routes began climbing. And climbing. In fact, they not only returned to pre–Reno Air levels but surpassed them. According to Oster and Strong, Northwest's average fare between Minneapolis and Reno in April 1993 was under $100; within six years the ticket prices ranged from $345 to $1,476.

Overall, the DOT received thirty-two complaints of unfair competitive conduct from new-entrant airlines between 1993 and 1999. Then the Transportation Research Board convened a panel of experts—including Alfred Kahn, the Father of Deregulation—to examine this phenomenon and concluded four major airlines were still engaging in practices to drive out new competitors.[3] The research board also defined "predatory pricing" as actions designed to drive out or suppress competition with the intention of later increasing prices.[4]

Back in 1983, *Time* trumpeted "Dirty Tricks in Dallas," detailing the Justice Department's federal suit against American Airlines and a taped conversation in which CEO Bob Crandall spoke to Braniff CEO Howard Putnam about both carriers raising fares. Other allegations, according to *Time,* included American pilots causing delays on runways to disrupt Braniff flights and American taking its time delivering $9 million for interline ticket agreements. The Harvard case study also cited ticket agents encouraging customers to fly American instead of Braniff and fabricating technical problems with American's aircraft so Braniff made costly plans by scheduling additional flights to accommodate passengers who never showed.

Another weapon in the majors' arsenal is their grandfathered right to operate at overcrowded, high-density airports where they charge higher fares, such as LaGuardia, JFK, O'Hare, and Washington National. Because major airlines control most of the takeoff and landing slots, they can use these rights to their advantage, even swapping them among one another, as Delta and US Airways did in 2011.

The Dirty Secret of Airfares: Bias

In the 1980s the computer reservations system Sabre, sister company to American Airlines, was accused of blatantly biasing screens for travel agents so American's competitors were listed below it. (An earlier study by American found that travel agents overwhelmingly booked the first carrier listed and rarely shopped on a second screen, a practice that still holds today among consumers surfing travel sites.) Former CEO Crandall tried to defend such practices before Congress by saying, "The preferential display of our flights, and the corresponding increase in our market share, is the competitive raison d'être for having created the system in the first place." But not surprisingly, biased displays were banned.

Even so, airlines have found other ways to game the system. One method seems to be with third-party travel sites, so that biased displays were carried over from reservations systems to travel sites. Shortly after I became editor of *Consumer Reports Travel Letter* in 2000, I decided to test such rumors by repetitively searching for identical fares on competing sites in real time, and benchmarking the fares through analyst Bob Harrell, who had access to Sabre and could simultaneously retrieve computer reservations data. Our first test, in October 2000, examined Cheap Tickets, Expedia, Lowestfare, and Travelocity and found "disturbing evidence of bias," including advertised airlines dominating flight listings and bogus itineraries listed first. In June 2002 we tested again—this time with six sites, including Orbitz—and confirmed they all received some compensation from airlines. But then and now, it's difficult to know how such payments affect displays.

Starting in 1995, airlines began cutting base commissions to travel agencies, from 10 percent down to 0 percent. However, many airlines continue to pay TACOs, travel agency commission overrides. These secret agreements can be tied to an agency's total booking volume for an airline, an increase in such volume, or even its volume in conjunction with a rival airline's volume. The danger to consumers, as the Oster-Strong report stated, is that overrides are often not revealed to them, so travel agents have an incentive to withhold information on competing flights.[5] Both the General Accounting Office and the DOT's inspector general also criticized the practice.

In 2001, we decided to test this theory at *CRTL*. We transferred our apples-to-apples testing methodology and applied it to brick-and-mortar travel agencies. While contacting 840 agents throughout the country and requesting flight and fare information on routes with low-fare competition, in real time we benchmarked through Harrell's reservations system. The results: Only 51 percent of agents provided complete airline and pricing information upon first request, and only 63 percent provided it when asked a second time. Overall, 25 percent of the agents failed to mention all the low-fare alternatives, and 12 percent didn't mention them at all.

By Any Other Name: Codesharing Deception

In the mid-1990s, travel advocate Bruce Bishins filed a regulatory complaint against codesharing that stated: "Ours is the only industry which permits Coke to be poured into Pepsi bottles and still sold as Pepsi." And Dr Pepper as well.

Every air carrier has a two-letter code issued by the International Air Transport Association: AA for American, DL for Delta, etc. That code is coupled with a flight number. But in the case of codesharing, two—or more—airlines can sell tickets on a flight operated by a single carrier, and usually such agreements are reciprocal, so in other regions that carrier sells tickets on flights operated by its partner(s). Today codesharing is found in three prevalent forms: between mainline carriers and their regional partners; between domestic competitors, such as United and US Airways; and within international marketing partnerships, particularly the three global alliances—oneworld, SkyTeam, and Star Alliance.

Not surprisingly, codesharing began with a regional carrier. Way back in 1967 a Henson Airlines flight from Hagerstown, Maryland, was linked in Washington, D.C., to Allegheny Airlines.

Ironically, those companies that most benefit from codesharing today were among the early detractors. When Bob Crandall was chairman of American, he vocally criticized the practice and later called it "bad for the industry and worse for the public interest." Back in 1984 the *New York Times* reported that a spokesman for United had assailed codesharing by saying, "It misrepresents two airlines with different levels of service which are indicated as one airline. We think it misleads the public." Today, of course, United is one of the most aggressive proponents of codesharing and a member of Star Alliance.

So who benefits from codesharing and international alliances? Unsurprisingly, it's not the passengers, because the practice increases airfares as well as airline profits.

In 2010 I cooperated with the DOT Inspector General's Office on an audit of codesharing practices and how they affect consumers. One inspector told me the goal was to bring attention to such policies, so there are no gaps or opt-out opportunities. The inspector general also agreed with my recommendation to the FAAC that the Transportation Department's Monthly Air Travel Consumer Report be reorganized with regional airlines aligned beside their mainline partners, so therefore consumers could easily see how *all* domestic airlines perform in the rankings each month.

The Death of Antitrust

For the legacy airlines, codesharing means big profits with little exertion, and the highest yields are found in international alliances. For example, oneworld reportedly generates more than $3 billion annually in "incremental revenues"—and that's the smallest of the three global alliances. Together the airline members of those three partnerships carry 73 percent of all passengers worldwide.

If you're an airline executive, there is virtually no downside to codesharing. As airline analysts note, it increases an airline's size and scope, which is critical. However, it's artificial expansion, in name only; Coke is still not Pepsi and United is still not US Airways, no matter how many marketing and advertising campaigns imply otherwise.

The long-standing antitrust laws were designed to keep airline executives from colluding on pricing. Of course, such laws have been broken repeatedly through the years. But antitrust immunity means illegal behavior is now legal, and price-fixing can be done in the open. The game-changer in airline codesharing came in January 1993, when Northwest and KLM were granted immunity for routes between the United States and Europe. Since then, dozens of other carriers have received such blessings.

A white paper published by Daniel M. Kasper and Darin Lee concluded: "Based on the empirical evidence, it seems clear that global alliances and antitrust immunity have produced to date enormous benefits for consumers, both in terms of lower fares and more convenient, integrated service." But the dangers to consumers are still emerging.

Not only do global alliances provide more market power that can be used against smaller competitors, but experts worry these partnerships can allow member airlines to bully airports and even governments. And the larger these alliances grow, the harder it will be to control them.

Also, there are dangers in picking the wrong partners when there are tangible safety or service issues. In June 2011 Delta found itself immersed in a public relations brouhaha when the

Huffington Post and other media organizations reported the SkyTeam alliance was welcoming Saudi Arabian Airlines, the flag carrier of a nation with discriminatory and anti-Semitic visa policies. In this case, codesharing meant Delta and other SkyTeam members were in the uncomfortable position of enforcing its partner's unsavory procedures.

The Peculiar Tone-Deafness of Airline Executives

"I honestly don't think many airline executives have any concept of what it's like to be an average passenger," says one industry veteran. It's a fair point. After all, senior execs don't get bumped. They board first. Flight delays and cancellations are rare indeed, since operations personnel ensure they are provided the best aircraft and freshest crews. Their baggage is not checked, let alone "mishandled." They never sit in middle seats or fight for overhead bin space. And they are never subjected to what I term the Information Vacuum—when passengers stare at flight information display screens and wonder, Will I get home tonight?

(For the record, I repeatedly contacted the current heads of several domestic airlines, but only Glenn Tilton of United/Continental and Gary Kelly of Southwest agreed to interviews; the CEOs of American, Delta, Frontier/Republic, JetBlue, Spirit, and US Airways did not make themselves available or did not respond in time.)

In July 2011 the *Atlantic* published data from the American Customer Satisfaction Index to compile "The 19 Most Hated Companies in America." Mixed into a list dominated by the electric, gas, and telecommunications industries were four domestic airlines: American (#8), United (#7), US Airways (#6), and Delta (#2). The comparison to commoditized utilities was not accidental. To many observers, airline executives work in a service industry but act as if they work for utilities, and don't seem to hear the rising crescendo of passenger complaints.

"They just don't get it," says Charlie Leocha of the Consumer Travel Alliance. "The reality is they treat people like cargo and they treat their workers the same way, as if they're numbers on a flowchart. They could have treated passengers better. They live in a different world."

4

So You Think You've Found the Lowest Fare

The airline business is the closest you can get to war in peacetime.

—*Herb Kelleher, chairman of Southwest Airlines*

WHEN THE AIRLINE INDUSTRY CLAIMS that airfares have remained stagnant over the last thirty years, my response is—well, yes and no. Fares are a function of competition, and on routes with no low-cost carriers, passengers get gouged. And one academic found fares fluctuate by the distance of the flight.[1] In a white paper on airline deregulation, Elizabeth Bailey of the Wharton School noted: "There has been enormous price dispersion from deregulation both within and across routes. Across routes, fares have fallen more on long routes than short routes." What's more, there is such a range of prices being offered on any given flight, it's entirely possible two passengers crammed in elbow to elbow paid fares that differ by a factor of ten. And let's not forget that added fees represent airfare hikes in disguise.

Many airline executives and Wall Street analysts, of course, look at airfares from a very different perspective. "They can't raise prices fast enough," says Helane Becker. "Our biggest concern is smaller airlines will get cocky and start to cut fares."

I spent thousands of waking hours from 2003 to 2006 endlessly trolling the Internet for the lowest airfares. As the project manager for a grant-funded analysis of online travel for Consumers Union, I went overseas and even examined travel sites in countries throughout Europe. Yet every time I was interviewed on TV or radio, my sound bite was the same: "Smokey Robinson was right—you gotta shop around." Our findings were conclusive: no single travel site offers the lowest fares on all occasions, and in turn the lowest fare could turn up on any given site. The complexities of airline pricing and airline distribution have created a labyrinth that no one can truly penetrate.

What's more, it can pay to shop at various times (particularly on Tuesday afternoons, when new fares are loaded into reservations systems) and to tweak your itineraries; departing even one day earlier or later can mean big savings. Using nearby airports can help land a better ticket price as well. And no matter where you find a good airfare, make sure you check with that carrier's own "branded" website, since the airline may be offering the best deal there.

But the flip side is that airlines watch their customers very carefully, and now more than ever, all customers are not created equal. I was on board a Virgin America Airbus A320 recently and the flight attendants repeatedly reminded us the forward lavatory was for the use of first-class passengers only; in other words, 8 people shared one toilet, while 141 shared the remaining two. So it's no surprise that ratio of 8 to 70.5 seems fair to the airline's pricing department.

It's a given that passengers in business class and first class are treated better—boarding first, more overhead bin space, a powerful curtain separating them from *us*. And naturally frequent flyer elite members travel in rarefied air as well. News reports in the summer of 2011 focused on Tom Stuker, the first person to reach ten million miles on United Airlines; the carrier not only named a Boeing 747-400 in his honor, but he even has his own dedicated reservations line (presumably not in India).

What many passengers don't realize is how the airlines further divide the rest of us in economy—by the type of ticket we purchase, how much we pay for it, and where we buy it. Sometimes this divide is all but acknowledged by an airline executive. For instance, after *Consumer Reports* published an airline survey in 2011 that ranked US Airways last, Robert Isom, the airline's executive vice president, stated: "I wonder if this survey really captures the customers we focus on." The flip side, clearly, is Isom acknowledging his airline carries an awful lot of customers US Airways does *not* focus on.

Don't let anyone tell you the booking channel doesn't matter—for the airlines, the medium truly is the message, and the message is all about how you'll be treated. Travel ombudsman Linda Burbank notes this as well: "I get very few letters from elite frequent flyer members." In fact, where you book and how much you pay for a ticket does have an effect. Burbank sees a lot of complaints from passengers who booked through third-party travel sites rather than directly through the airlines. This can lead to issues over fare increases and change fees, since now you're navigating the fine print produced by two sets of corporate attorneys.

In the case of business travelers, their tickets are often bought through corporate travel departments or mega-travel agencies that receive discounts and other perks based on volume. Again, not all passengers are treated equally. Reservations agents are often paid based on productivity—that is, tickets sold—so spending time on the phone assisting passengers who have already booked actually costs the employee money. Little surprise that some agents choose to "transfer the call to limbo."

In 2011 I attended a conference where Pauline Frommer, daughter of travel writing icon Arthur Frommer and editor of Pauline Frommer Guides, said, "If you book ten months before you travel, you're a sucker." So when should you book? Oh, boy.

Sometimes bargains are found early in the process, and sometimes quite late. How's that for a wishy-washy response to a dysfunctional industry? "It's incredibly frustrating," admits Jami Counter of TripAdvisor. "If you're comfortable with the price, you almost should just buy it."

The alternative is to keep shopping. But at least airline pricing has gotten simpler thanks to the Internet, right? No one's ever uncertain about a good airfare, huh? Of course, the guy next to you may have paid $400 more, or $1,200 less. And an upgrade to business or first may be four to six times the cost of economy, though you do get that $3,500 hot towel. Just choose the booking channel and you're on your way. Visit an airline ticket office? No, most have been shut down. Call? Don't forget that fee. Contact a travel agent? The few not driven out of business by airline commission cuts don't usually find it profitable to process tickets anymore. Go online? Sure. But where? Expedia, Orbitz, or Travelocity? An "opaque" blind bidding site like Priceline or Hotwire that conceals the name of the airline and the itinerary details until after your credit card is charged? A wholesaler or consolidator that offers no refunds? The airline's own "branded" site? Don't forget, you still need to factor in Saturday night stayovers. And the time of day, week, and year you'll be flying. And peak and nonpeak travel times. And an advance purchase window of 7, 14, or 21 days. And whether or not it's refundable. Or cheaper at a nearby "secondary" airport. You'll probably need to connect at a godforsaken hub located eight hundred miles in the wrong direction, so add on some additional fees. Of course, you need to calculate the mandatory add-ons, including federal taxes, passenger facilities charges, 9/11 security fees, and fuel surcharges. And then you can start adding up the rest of the fees. Otherwise, yes, shopping for an airline seat has never been easier.

The Complex World of Airfares

According to one of the mega-travel agencies, at any given moment there are *six billion* airfares housed in the industry's global distribution systems, formerly known as computer reservations systems. About half of those fares are designated for corporations, leaving roughly three billion different pricing combinations for everyone else. No wonder most people think the guy in 24F paid less—chances are, he did.

A lot of what confuses, frustrates, and angers airline customers can be traced back to simple letters of the alphabet: those "fare basis codes" that are stamped onto the tickets. The most recognizable are F (first), C (business), and Y (economy). In fact, each airline can have a dozen or more subclasses of fares, with as many as twenty-six separate airfare codes for one class of service.

These letters indicate the many types of restrictions that may apply to each ticket, such as advance purchase (when the ticket was bought), minimum stay requirements, and refundability. Like snowflakes, no two flights are priced exactly the same, and complex yield management logarithms determine the mix of fares available based on routing, time of year, day of week, time of day, etc.

There's a flip side to this complex ranking of passengers. Not only do the codes indicate how much a customer pays, but they also indicate how that customer will be treated during "irregular operations." The ticket price and the booking channel will determine if the passenger is bumped or encounters an extended delay or cancellation.

The Associated Press summed it up: "The good news: The better code you have, the better your chance of not getting bumped. You also might receive more frequent flier miles if you're in the top tiers. The bad news: The main way to improve your code is to pay more." And woe to those who purchased their tickets online, but not through the airlines' own "branded" sites; consider that customers who buy tickets through "opaque" sites such as Priceline and Hotwire forfeit frequent flyer mileage entirely.

It wasn't always like this. Traditionally there were only four fare categories: F and Y for first and coach, FN and YN for night flights. And the same fares applied from entire regions of the country, so prices were identical from cities such as Boston, New York, Philadelphia, or Washington to cities such as San Diego, Los Angeles, or San Francisco.

Airfare pricing expert George Hobica also knows of no other industry other than perhaps semiconductors with such complex pricing: "I don't think you can talk about it in rational terms. It's a race to the bottom." There's that term again, only here applied to pricing rather than cost structures.

Analyst Bob Harrell can even pinpoint the date airline pricing changed for good in America: April 24, 1977, when American Airlines introduced the SuperSaver program, offering discounts of up to 45 percent off fares between New York and California, which the carrier describes as "the most popular fare in its history." Note that this occurred prior to deregulation in 1978, as the Civil Aeronautics Board under Alfred Kahn was ushering in a new era. Harrell says, "That was the beginning of the incredible level of complexity. Pricing complexity goes up and down. They're always tweaking it. But most of the scientific ways to bring more revenue bring more complexity."

For consumers, the antidote in recent years has been low-cost carriers such as Southwest, which offer a more simplified fare structure, often consisting of just several prices for a single itinerary. But even bare-bones Southwest has allowed complexity to creep into its system, particularly after introducing fully refundable Business Select fares.

Time and again, I've run into this issue, when people ask me that simple question—"Is this a good airfare?"—and I'm forced to sigh and respond, "Well, you see in 1978 the industry was deregulated. . . ."

The Commoditization of an Airline Seat

In researching this book, the one term that was most often raised by others without prompting was *commoditization.* Analysts note there's little difference among economy seats these days, so a plane is a plane is a plane.[2]

Think of it in terms of hotels. There is clear differentiation among brands—no one confuses Econo Lodge and Ritz-Carlton—and that carries over into how travel sites display these products. There is much arguing within the lodging industry over how hotels are rated by stars, but even allowing for some latitude in such subjective rankings, the consumer is never confronted by a booking screen that lumps in Super 8 and Four Seasons side by side. So why don't customers choose a hotel room strictly on price, as they do with airlines? I once had a hotelier explain it to me this way: "There's a big difference between which seat you'll rest your ass on and which pillowcase you'll rest your head on."

Yet airline executives argue their brands are different, too—how can you compare full-service United and low-cost Southwest? But every day hundreds of thousands of travelers do just that: compare them side by side strictly on fares. What's more, Southwest consistently ranks higher in customer satisfaction than United, which industry wisdom says is because Southwest does a better job of "managing expectations."

Industry experts note the Internet has kept travel prices down by pressuring airlines and other travel companies not to be ranked unfavorably.[3] In other words, no airline wants its lowest fare to be listed in the Siberian confines of a second screen, particularly since marketing research indicates many consumers never scroll down that far anyway. So the Web has become the great equalizer, bottom-lining every airline's lowest price even if some of those airlines offer very different products, services, and cost structures.

Actually, Rolfe Shellenberger traces commoditization back to the airline industry's roots. He points out that in the preautomation days, the most critical travel publication was the *Official Airline Guide*, a phone-book-sized text first launched in 1929 that listed the times and destinations of every commercial flight in the country by carrier. In those days passengers generally selected flights by choosing an airline, since price was nonnegotiable and flight frequencies were often duplicated. Then in 1958—the very same year the jet age began in earnest with the introduction of the Boeing 707—the *OAG North American Quick Reference Edition* was launched, which sorted and sequenced the flight schedules of all airlines. Thus an entirely new way of booking flights became available. So for the first time consumers could look up flights based on routes and time of day, and not just by airline. As Shellenberger says, "The *OAG* bastardized the entire industry by saying all airlines are the same."

The bad news for consumers today is much less competition for passengers based on service, comfort, or amenities. Michael Levine tells the tale from his airline days: "Anytime we experimented with the idea—we'll give them something a little better, but it will cost them a little more—they turned it down. And the purest forms of these experiments were run by American and TWA when they took seats out for more legroom programs." Despite the clear advantage of these two programs, neither carrier made any money from offering more comfortable seats. Levine concludes: "Of course, we in the industry watched this with great interest. But the evidence is, there is something in the consumer mind that says I'm only going to be on the airplane for two hours, four hours, twelve hours, fifteen hours—and I'll put up with whatever it takes to save a few dollars."

Many believe that commoditization in large part is driving the race to the bottom on everything from service to safety. Labor leader Pat Friend is not afraid to raise a corollary issue: "You hear consumers complain about getting bad service. I can't tell you how frustrating it is as a flight attendant to not have the tools to serve the customers. I tell my friends, when you stop surfing the Net for the cheapest fare you can find, then we'll talk about service."

George Hobica and his colleagues at Airfarewatchdog.com spend hours every day surfing for airfare bargains, and he says that the irony is that some low-fare airlines are not even competing on flights, but on other products they are paid to sell online to customers booking vacations: "Allegiant Airlines doesn't make money on those low fares, it makes money selling tickets to Blue Man Group." Hobica is right: in an examination of forty-seven carriers worldwide, Allegiant topped them all by notching 29.2 percent of its total revenue through ancillary fees; in second place was Spirit at 22.6 percent and in third was Ryanair at 22.6 percent. The only legacy airline on the topten list was United/Continental at 14.7 percent. As IdeaWorks stated: "Many of these airlines are becoming savvy retailers." Lesson learned: for the airlines, getting you to shop on their own branded sites is critical.

Experts point to the world's undisputed low-fare leader—Ireland's Ryanair—and its mercurial CEO, Michael O'Leary, who continually makes headlines by promising airborne pay toilets and "stand-up" fares without seats. It was Ryanair that dared to impose baggage fees and made them stick, a move that was first emulated in the United States by Spirit.

That Dirty Little Word: *Fee*

Like all Americans, I encounter ancillary revenue in all sorts of places. My bank recently docked me a fifteen-dollar "maintenance fee"; my gas station charges more for noncash purchases; and my apartment complex introduced a "Preferred Parking Program" by roping off the same old spaces and charging an additional twenty-five dollars per month (after midnight, tow trucks ensure compliance). But no other business can match the airline industry for sheer artistry in levying fees.

Ancillary revenue. It's quite a term. A fine example of corporatese—specific enough to be understood in dog-whistle fashion by a select few, but vague enough to be unrecognizable by most of the population. So don't expect the airline industry to call ancillary revenue by more apt descriptions. Added fees. Nickel-and-diming. Gouging.

Dozens of airline executives are cursing me for that last paragraph. They'll patiently explain why "unbundling costs" is a necessity, a happy medium between raising airfares and filing for bankruptcy. But one thing absolutely cannot be denied: the U.S. airline industry has done a terrible job of communicating such needs to its customers.

According to the DOT, U.S. carriers raked in $3.4 billion in baggage fees and $2.3 billion in reservation change fees in 2010. But here's the kicker: that $5.7 billion is only part of the story, since the DOT cannot separately identify other types of fees lumped in with ticket revenues.

Still, the industry has its defenders. "The whole thing pisses me off," says Brett Snyder of CrankyFlier.com. "Not that there are fees, but how the public perceives them. . . . Airlines have not done a good job of merchandising their products. People have these rose-colored glasses about how glamorous flying was before deregulation. Fees are a way to keep their heads above water." On the other hand, even Snyder acknowledges some carriers go too far, especially the "convenience fee" of $14 with Allegiant Air and the "passenger usage fee" of $4.90 each way with Spirit Airlines—to book *online*!

There has been no stronger hot button among passengers than added charges. But Michael Levine makes the case for some of these fees: "You charge more for the window seat so you can charge less for the middle seat. Your total revenue will be the same. . . . The customer will say, I don't see why the window seat costs more than the middle seat. But by the way, they have no problem walking into a theater and paying more to sit in seventh row center than in the second balcony. It's the same goddamn thing. The theater is the same size every night. It costs the same to put on the play every night."

Glenn Tilton doesn't just defend baggage fees—the chairman of United/Continental goes further by saying he doesn't understand why the airlines *ever* allowed bags to be checked for free: "My first question when I joined the company was, how much investment do we have tied up in bags? The answer was more than a billion dollars. Then I asked how much it brings in and the answer was, it's negative. So I asked, why don't we charge for bags? I got nothing but blank stares. And every day we were bleeding five to seven million dollars." Tilton can discuss the macroeconomics of the airline business and the need for ancillary revenue for minutes at a time, but in the end he agrees with me on one point: the industry has conducted a poor public relations campaign in explaining such fees to consumers. At a bare minimum, airline executives should endeavor to explain to customers what they explain to board members; if fees are so necessary to a company's financial health, then why should it be so hard to make consumers understand?

When the Bottom-Line Fare Isn't

Ancillary fees now constitute about 5 percent of the industry's total revenue, and for the last few years a small war has raged within the travel industry over full disclosure of such fees, particularly for corporate travel departments that have seen their budgets skyrocket. This has affected average consumers as well, even those traveling for pleasure. Some 60 percent of consumers buy airline tickets through sales channels where airlines don't provide all the data on optional services fees. In other words, customers can't obtain the bottom-line cost of an airline ticket, an otherwise sacrosanct consumer right in all other areas of American commerce.

In fact, many domestic airlines currently are in violation of pricing transparency regulations, because if a customer cannot get the total price in one click, it's a DOT violation.[4] That's why Kevin Mitchell says, "The majors would argue, 'These fees prevent fare increases.' But in my heart I feel these guys are double-dipping."

The Holdouts on Bag Fees: Southwest and JetBlue

Jay Sorensen of consulting firm IdeaWorks believes airline ancillary revenue will "top" at about 30 percent of total airline revenue, which certainly seems to translate into a fare hike for most travelers, if a $200 fare becomes $260 with added fees. He also says charging extra for checked bags will become ubiquitous worldwide over the next two years, in part because of international codesharing and marketing alliances.

But with bag fees, there are two notable exceptions. For no

matter how hard airline public relations execs spin on this topic, two irrefutable facts have emerged: the first is that Southwest and JetBlue are the only two U.S. airlines not charging for first checked bags and the second is that Southwest and JetBlue are the two most popular airlines in the country.

While JetBlue does not charge for first bags only, the Southwest "Bags Fly Free" slogan applies to both the first *and* second checked bags. And it's not just about surveys and awards—those good vibes are translating into added business in an industry notorious for a lack of brand loyalty: Southwest has notched market share advances since baggage fees were introduced.

So, will Southwest hold firm and buck the baggage fee trend? Some believe that if fuel gets any more expensive than it is now, Southwest will have no choice but to start charging baggage fees. Others think that because Southwest is so invested in its status as the low-cost airline, it will have to maintain its current policy. I asked CEO Gary Kelly if his airline will continue its bag policy, and he responded immediately: "I guarantee it."

But Southwest's cost structure was aided in large measure by the long-term practice of "fuel hedging," which locked in a set price for a considerable and volatile expense. It's a dark art that requires guessing right *and* having cash collateral. For 2008, the *Los Angeles Times* reported that about 70 percent of Southwest's fuel that year was purchased at approximately $51 per barrel, at a time when crude was pricing at $126 per barrel. While its competitors were reeling from fuel costs, it was a golden opportunity for Southwest to hike airfares or impose fees, which in turn would have helped other airlines do the same. Instead, Southwest did neither.

In 2008, when Southwest was heavily invested in fuel, it declined to raise its fares. And by not raising ticket prices, Southwest helped force other airlines to impose fees instead—and then Southwest declined to match those fees, gaining further advantage over its rivals. Southwest passengers won—twice.

A report in February 2011 noted that United/Continental imposed fuel surcharges, yet another add-on fee not always captured in the bottom-line price of a ticket. In turn, this caused price hikes from American, Delta, US Airways, Alaska, and AirTran—but not Southwest. Pricing analyst Bob Harrell posits an important theory: "I think there's a lot of whining about fuel costs. Fuel cost fluctuations have been around for thirty years now, but airlines use it as a way to justify fare increases. So I think airlines like the price of fuel going up. It improves their margins. The increased charges go in immediately but they come out slowly."

In 1993, the DOT issued a landmark study titled "The Southwest Effect," which demonstrated how the nation's low-fare leader gains market share, drives down fares, and generates new air traffic in the cities it enters. The study also included this key observation: "Southwest's demonstrated ability to quickly dominate markets and force out competitors may not be perceived as a problem in the near term because Southwest offers lower prices, even as a monopolist, than other major airlines offer even in the most competitive markets." But now more and more industry experts are contending there's been a change, and I hear it over and over. George Hobica sums it up: "With Southwest, the game is over. They're not as low-fare anymore." Even the American Antitrust Institute noted the "Southwest Effect" has not always occurred in recent years.[5]

CEO Gary Kelly doesn't deny this, and notes, "Our cost advantage has narrowed." But he says fares are tied in with costs, and as Southwest's costs lower, so will its fares.

Bob Harrell believes the overall demarcation between low-cost carriers and legacy airlines is becoming more obscure: "The lines are definitely blurring. The LCCs are trying to maximize their profits and the legacies are pushing back. They used to just run away from Southwest but there's only so much geography in the country." And the same can be said of JetBlue, which some claim has been "coasting" on a low-fare reputation.

How the Southwest Effect affects the rest of us is critical: if America's leading low-cost airline is not so low cost anymore, the fallout on airfares will be tremendous in coming years. But columnist Joe Brancatelli bucks the conventional wisdom when it comes to Southwest: "Everybody's always looking for the trap-door. But thirty-six years later, they're still making money. I think [CEO Gary] Kelly gets the benefit of the doubt." Point taken.

Frequent Flyer Programs: Rewards or Ponzi Schemes?

The introduction of frequent flyer programs injected just a pinch of long-lost romance into the airline experience. Sure, it sucks to be squeezed into a middle seat, swiping a credit card for a Sprite Zero. But . . . a quick calculation . . . maybe . . . let's run the numbers . . . just maybe . . . Cancun! Victory!! Who didn't root for George Clooney in *Up in the Air*? (It's worth noting that in Walter Kirn's novel it was the fictional Great West Airlines that was featured, but in Ivan Reitman's film the brand placement of American Airlines was so heavy the company seemed like a character unto itself.)

As with other pursuits, mileage expert Randy Petersen says everyone wants to earn low and redeem high. But those who entered such programs in the early days have done much better than those who have entered more recently. And that's a model that really doesn't resemble the stock market. It actually resembles a Ponzi scheme.

Consider that just a few years ago—before expiration rules tightened—there were more than 10 trillion outstanding miles, while members were redeeming less than 900 billion annually.

And further consider that today more than half of all frequent flyer miles are generated without ever boarding an airplane. That's a staggering factoid for programs that were touted as "loyalty" and "rewards" plans to steer passengers to Carrier A rather than Carrier B.

The statistics have become more elusive, but experts estimate about 9–10 percent of all airline seats are occupied through redemption. Linda Burbank, the travel ombudsman, says she no longer sees airlines offering frequent flyer mileage as compensation in resolving customer service disputes, which undoubtedly is a reflection of the growing percentage of unredeemed miles on airline balance sheets.

The problem for airlines and passengers alike, of course, is the same problem discussed in other chapters: record-high load factors. With planes more crowded than ever, the number of unredeemed miles keeps soaring. So in order to mitigate a clear-cut case of demand far outpacing supply, the major airlines have developed other coping mechanisms, like higher redemption thresholds, expiration dates, and fees, fees, and more fees.

The bottom line is that the goalposts have been moved. Until a few years ago, it required 25,000 miles to earn one round-trip domestic flight; now all the majors require 50,000 miles in most cases. Of course, redemption is still available at 25,000 miles, but it includes a host of restrictions and blackout dates that affect availability.

Then there are the ever-changing rules on expiration. As far back as the mid-1980s, United was stamping expiration dates on mileage; today it's common practice. For example, Delta announced that all mileage earned after January 1, 2011, will not expire, but all mileage set to expire prior to that date would not be reactivated on a complimentary basis. I received a letter from an angry reader who was told by a Delta rep he could reactivate his expired miles—for a fee (there's that word again). He rightfully asked why he needed to pay for something he had already paid for once.

And of course there's no reason to believe frequent flyer programs will be spared the ancillary revenue trend. SmarterTravel.com's "Ultimate Guide to Frequent Flyer Fees" is continually updated, and no wonder, since both the number and amounts of fees are continually increasing. An "Award Ticket Change Fee" is $150 on American, Delta, and US Airways, while "Upgrade Cash Surcharges" can be as much as $500 on Continental and United, and Alaska charges $25 for something called a "Partner Airline Award Fee."

Many members may not have contemplated it, but in essence frequent flyer programs are a study in comparing apples to oranges. On the flying side, you earn by generating miles, so a flight from Boston to Seattle is worth about three times a flight from Boston to Detroit. Yet once it comes time to cash in those miles, all domestic round-trips are treated equally, so Boston–Detroit requires the same 50,000 miles as Boston–Seattle. What's more, airline pricing has never been based on mileage; you can easily pay four times as much to fly from Boston to Washington as from Boston to Miami. So there really are three types of currency—miles, trips, and fares.[6]

Therefore, when it comes to redeeming, consumers should consider the *value* of a single frequent flyer mile. Like all currency exchange rates, it's a moving target, but it usually hovers in the 1.5¢ to 2¢ range. That means even a trip that costs only 25,000 miles should be redeemed only if it cannot be booked for less than $400. The airlines will never tell you this, because they're happy when you redeem mileage at less than its face value—whether it's for trips, upgrades, or any number of products in that catalog in the seatback pocket.

Tim Winship is another guru, as well as the publisher of

FrequentFlier.com, and he speaks of two "marquee" pieces of news in recent years: One is the evolution from frequent flyer programs to frequent buyer programs. American Airlines' AAdvantage program has grown to more than one thousand partners, for everything from financing a mortgage to buying a pair of khakis at the Gap. So how can an airline executive resist when Citibank comes calling with a satchel full of cash to cobrand a credit card? Winship says it would be bad business to say no. The other piece of news is that awards are increasingly difficult to come by.[7] Sorensen of IdeaWorks adds, "I think the airlines have learned the economic cost of ignoring frequent flyer members because of how it has come back to bite them."

Winship points to the industry's recent "historic" escalation in filling cabins while both load factors and the amount of miles being issued continue to increase: "It's a zero sum gain." So at what point do we declare airline frequent flyer programs to be well-marketed pyramid schemes or shell games? "That's the worst-case scenario," admits Winship. "Would the airlines allow this to happen? You never want to underestimate the stupidity of the airlines. My own best guess is they would avert that type of shipwreck."

Any discussion of frequent flyer programs eventually leads back to Rolfe Shellenberger. Back in 1951, he was one of about one hundred people on the second floor of Hangar 3 at New York's LaGuardia Airport, answering calls and jotting down passenger names and phone numbers on index cards, cutting-edge Korean War–era technology. Eventually Shellenberger would become a key architect of AAdvantage, widely recognized as the first major mileage program when launched in 1981. Airline myth holds that launching frequent flyer programs in the 1980s was never about rewarding loyalty—it was about *buying* loyalty in an industry notorious for a lack of it.

As for Shellenberger, he is not happy to see how his brainchild has devolved. "It's cupidity," he says of the current state of frequent flyer programs. "It's greed. They're screwing it up and they're ruining a good thing. The bean counters got involved because it's profitable, and few things are profitable for the airlines these days."

5

A Mad Race to the Bottom: How Airlines Mistreat Employees, Outsource, and Ignore Passengers

Once you consent to some concession, you can never cancel it and put things back the way they were.

—*Howard Hughes, chairman of Hughes Aircraft and TWA*

NO PUERTO RICANS, NO JEWS, no job.

In 1991, I learned that my employer, the Pan Am Shuttle, had been sold to Delta and would commence operations as the Delta Shuttle. During the course of a hectic twelve days, I was interviewed by the conquering carrier and offered a job 750 miles away. A senior exec from Delta personally sat me down in the Marine Air Terminal at LaGuardia Airport and urged me to jump at employment as a flight dispatcher in Atlanta. But I wasn't eager to jump; in my heart I knew my "summer" sojourn working for the airlines—which had now lasted seven summers—was finally ending. I had no interest in working in a large ugly building without windows at Hartsfield-Jackson International Airport.

I patiently explained I had personal reasons for not transferring to ATL: I had been married for only two years, and my then-wife was immersed in a Ph.D. program and could not leave the area. It would mean living in a one-room efficiency in Peachtree City and using my dispatch license to bum cockpit rides back and forth to New York. The Delta exec tried to convince me my wife should switch schools: "We've got lots of good universities in Atlanta." Then he looked right at me and added: "In fact, about the only things New York has more of than Atlanta are Puerto Ricans and Jews."

When each of us looks back on the highlight reels of our life, we tend to focus on the moments when we stepped up and rose to the challenge. To be blunt, I've always considered myself fairly quick with a quip or a comeback, someone who can think on his feet without getting tongue-tied. But that day in the Marine Air Terminal I was flabbergasted. I simply couldn't conceive that a senior executive at a Fortune 500 company would utter such a statement, and my vocal cords failed me. For several long seconds, I had nothing. By the time I was close to retorting, he was blabbing about Delta's health benefits. I felt my face and neck growing warmer, and thought of some way to let this bigot know that the woman he wanted to switch graduate schools was Jewish. Finally I blurted out, "Another reason I can't relocate right now is my wife has a part-time job teaching Hebrew school."

You'd think he would have had the grace to look shocked, but he didn't even blink. Later that day, I learned from a good friend of mine that her interview had taken a bizarre turn as well. Her family was Cuban, and the man from Delta wanted to know where she was from. "New York," she replied. "No," he kept repeating. "Where are you *from*?"

It turns out our experiences were not at all isolated. Several weeks later, *Newsday* ran a front-page headline about Delta's treatment of former Pan Am employees that satirized the company's old slogan: "They Love to Pry and It Shows." There were multiple stories of applicants for the position of Delta flight attendant being asked about their private lives. Homophobia and sexism abounded. Males were questioned on their sexual orientation, while females were interrogated about birth control. By the early 1990s it eventually became apparent that Delta's definition of corporate diversity entailed promoting straight white Christian guys who hailed not just from Atlanta but from throughout the entire state of Georgia.

Then again, bigotry was not a new concept at Delta. Back in the day, Carleton Putnam had orchestrated the mergers of a series of smaller airlines that eventually became Delta, and then served as chairman. But in the 1950s he stepped down to devote himself to academic discourse, and promptly wrote *Race and Reason*, a landmark 1961 work on segregation. That is, it's a landmark work if you see the world through the eye slits of a hood made out of bedsheets. It's a book that *defended* segregation, and it's still quoted today by white-supremacy organizations. No less a literary scholar than David Duke—American Nazi and former Grand Wizard of the KKK—gushed that when he bought it, "I was about to read a book that would change my life."

So clearly there was a lot more playing out here than the melding of two very different corporate cultures. Pan Am was the world's greatest airline and America's premier flag carrier during much of the American Century, making its reputation at a time when Delta was better known for crop dusting. Human Resources liked to tell us our "blue ball" was the second most recognizable logo in the world behind Coca-Cola's circular thingie. And that blue ball was bolted into the top of a skyscraper that fronted Park Avenue. For my interviewer and several other Delta execs, the takeover of Pan Am's shuttle and European routes apparently was a response to General Sherman's march through Georgia. A few days after my Puerto Rican–Jewish interview, I was alerted to a modest severance package. That option was a hell of a lot more attractive than debating the merits of

Race and Reason on a midnight shift in Georgia, so I said farewell to a business that was rapidly changing—for the worse.

Goodbye to All That

My most controversial action on the DOT's Future of Aviation Advisory Committee was vocally and repeatedly arguing that the FAA needs more resources to address airline maintenance outsourcing, both domestically and overseas. There were three labor representatives on the FAAC and I was the only consumer advocate, but we at Consumers Union made a conscious decision not to address outsourcing as a labor issue, for fear it would dilute and distract from the very real concerns over safety. But the time has come to address what I could not address on the FAAC: there is no question that employee "downsizing" and outsourcing have had adverse effects not only on safety but also on service and quality.

In recent years most of us have become our own grocery clerks, bank tellers, gas station attendants—and reservations agents. Like other global industries, airlines have shifted tasks that once were performed by their own employees onto their customers. When we book online we become res agents, and when we check in at an airport kiosk we become ticket agents. And yet from an airline cost point of view, enough is never enough. Besides fuel, labor remains the highest debit on every carrier's ledger, and airlines keep looking for ways to reduce employee expenses.

Consider that in the 1990s and 2000s four domestic network carriers launched six different "airline-within-an-airline" products, separate brands designed to compete with Southwest and other low-cost operators:

- Continental Lite (aka CALite), from Continental
- Delta Express and Song, both from Delta
- Shuttle by United and TED, both from United
- MetroJet, from US Airways

All six efforts ultimately failed, and analysts noted that just wishing for a lower cost structure doesn't automatically produce one. These operations were run exactly as the high-paid consultants advised: no frills, skimp on amenities and meals, quick-turn the airplanes, obtain greater utilization of aircraft and equipment, etc. But there was no getting around the most significant variable cost of all: labor. Even when network airlines try to segregate their employees through separate contracts and hiring practices, legacy carriers manage to generate higher labor expenses.

And so the major airlines—and even low-cost carriers—have turned to other methods to reduce their labor bills. These include outsourcing, laying off workers by the tens of thousands, hiring "b-scale" employees at lower wages, shifting full-time work to part-time work, and cutting benefits.

What is not readily recognized by the airlines is that all these actions have equal and opposite reactions felt by passengers.

India Calling: Overseas Reservations

Airline customer service—increasingly viewed as an oxymoronic term by legions of paying passengers—has morphed into something quite strange in recent years. And so I decided to tap into the very nerve center of the U.S. airline industry. That, of course, meant traveling 7,834 miles across the world to India.

In the years after 9/11, news outlets across the country were continually reporting on U.S. airlines shutting down telephone call centers. In most cases these stories were accompanied by corollaries detailing where all that work went. In 2003: Delta outsources to India. In 2004: American outsources to Mexico. That same year: United outsources to Canada. Manila. Santiago. Cape Town. Montego Bay. Pune, India. Delhi, India. India. India. India.

Was anyone truly surprised when CBS News reported that eighteen states were outsourcing welfare benefit calls to India? One IT firm in India boasts on its site of all the Fortune 500 companies that have offshored work to that country, including Microsoft, Oracle, Citibank, Morgan Stanley, Wal-Mart, AT&T, General Electric, Reebok, General Motors, Sony, Boeing, Coca-Cola, Pepsi, Swissair, United Airlines, Philips, IBM, Lucas, and British Aerospace.

It was a delicate mission I embarked on, and as I precariously scheduled meetings at leading call centers in New Delhi and Mumbai, I lived in dread that a representative would insert the search phrase *McGee airline outsourcing* into Google and read all my previous work on the topic. Full disclosure: even before I began this book project, I wasn't a public cheerleader for offshoring U.S. aviation jobs.

Malarone for malaria and Vivotif for typhoid? Check. Deet insect repellent for warding off disease-carrying mosquitoes? Check. Compression socks for avoiding deep-vein thrombosis during the thirty-six hours of round-trip flying? Check. Visa? Check. But not before noting with irony that the Indian consulate in New York City has—wait for it—*outsourced* its visa services. LMAO, as the kids say.

The very week I was visiting call centers on the subcontinent, Donald Trump was making news, barking about America being overtaken by the new economic powerhouses, China and India. In one sense he's right, of course, since U.S. debt to Asia continues to accrue at an alarming rate; Southern California Public Radio reported last summer that China held at least $1.115 trillion of our debt. However, I've visited both countries and what Americans often don't grasp is that these emerging superpowers—complete with high-tech infrastructure, plenty of new construction, nuclear weaponry, even space programs—remain dirt-poor places for many of their own citizens to live and work. We have far too much poverty in America, of course, but there's no comparison between the standards of living.

Yes, Fortune 500 companies have rushed to outsource in India, but it's impolite for returning American business travelers to mention the piles of dung and the ubiquitous trash lining nearly every street in New Delhi and Mumbai. The cars and buses and Thai-style tuk-tuks clearly have no emissions controls, and the air stinks of pollutants, both gasoline-generated and animal-generated. When you set out for the airport at 5 a.m., packs of wild dogs roam downtown thoroughfares, cows graze on garbage in the traffic circles, and bare-butt children sleep in dirt. Sure, you'll see yuppies in golf shirts toting laptops. But once United Airlines and American Airlines and Travelocity come calling, it's no wonder there is no shortage of newly minted IT and customer service professionals eager to jump into the breach.

Of course, frontline employees on both ends of this equation are simultaneously getting screwed over on two continents. U.S. workers are watching decent jobs emigrate for good. James P. Hoffa of the Teamsters recalls meeting with devastated TWA employees who were in tears after their call center in Charleston, West Virginia, was outsourced to India. "Do you know what that does to a community?" he asks.

Meanwhile, in India they're barely making a fraction of the living wage they would receive in the Western world. Yet even where wages are lowest, there is little job security. Documentary filmmaker Dawn Mikkelson noted this when she visited outsourced maintenance shops in Asia: "The guy in Hong Kong is going to lose his job to someone in China, and now the guy in China is going to lose his job to somewhere else."

All this contributes to an economic divide in India that makes America look like a commune by comparison. It's crass to use the word *caste* in democratic India today, but how else to describe the creepiness of staying at the palatial Hilton in Delhi? It's a fenced, gated, and guarded fortress in which the undercarriage of every taxi is screened for explosives while the gorgeous fourth-floor swimming pool overlooks a slum. But it served as my base as I set out to finally view airline outsourcing up close and personal.

There are many entrances to the WNS Global Services facility in Mumbai, but my appointment was at Plant 10, Gate 4 (just past the Citibank kiosk at Gate 2). A poster warned NO SHOOTING! and they mean it—all cameras are checked at the door, and surreptitious photography results in confiscation. Once I turned over my digital Polaroid, I entered a huge campus that is quasi-collegiate and quasi-military. Then it hit me: the place resembled those old clips of 1940s Hollywood studios, with bicycle messengers scooting past Quonset huts.

The executive who scheduled my meeting had been called away, so instead I met Shainon Vyas, a hip young assistant manager in Corporate Communications (in apologizing for his delay he said, "My bad"). But he is Indian through and through, and repeatedly referred to Mumbai as Bombay. Unfortunately, nearly all my questions were deemed proprietary, so there was much he could not share with me. I received a quick tour, and gazed through thick glass at a vast room of endless cubicles designated exclusively for the travel sector. This was it: the epicenter. When you haggle for an airfare refund or need to reschedule your flight, chances are good your call has been routed across several oceans right to that desk over there.

For WNS, it all started back in 1996, with British Airways as the first airline customer. Since then the company has grown to more than 21,000 employees, with twenty-three centers throughout the world and annual net revenues of $616 million.

But even the corporate fact sheet doesn't list client names, just descriptions such as "major North American airline" and "North American travel agency."

India is a sweet spot for call center outsourcing, and the experts cite the combination of technological infrastructure and an endless employment base of educated and English-speaking recruits (WNS cattily notes the "language problem" in Malaysia, Singapore, and Thailand). But it's a hush-hush business, routing customers' calls to the far side of the world, and Vyas deferred on most of my key questions. What do your workers earn on average? Sorry, our clients wouldn't want us speaking about that. Well, who are all your travel clients? Sorry. Can I sit in while the travel team fields some calls? Sorry.

What he did say is this: "We want to remove this myth that this is not a primary career. There is substantial growth. We promise growth based on personal performance. We are an HR-oriented organization." In fact, Vyas did explode a few myths for me: WNS is not composed of an army of part-time workers; in fact, *all* employees are full-timers, though most are between seventeen and thirty-five. They work nine-hour shifts. And everyone handling travel calls—at locations throughout India as well as in the Philippines—receives full health benefits.

So if the staff is full-time and there are no slave labor conditions, how the hell can WNS boast on its site that it helped Travelocity reduce its operational costs by 40 percent year on year? "There is a cost advantage for sure," Vyas said in obvious understatement. So it begs the question, why have all those airline employees in Georgia and Oklahoma and Texas and Connecticut been fired and their jobs shipped to Mumbai, if WNS and other outsourced companies are such cool places to work? Where do those labor savings come from exactly?

The answer, naturally, is in the wages. Just a few weeks before I arrived in India, the PricewaterhouseCoopers branch in that country published "Measuring Human Capital—Driving Business Results," an extensive survey of thirty-seven Indian firms across various business sectors. The findings were not surprising: the "information technology and information technology enabled services" (IT/ITeS) sector recruits the most new employees each year. Yet IT/ITeS also has the lowest function cost per employee rates, as well as the highest termination and resignation rates. In other words, they get 'em in and then they get 'em out.

As for salary, the report notes Indian companies pay an average annual wage of about $10,733 in U.S. dollars, and earn a profit of about $13,417 per employee. But it's important to remember that this wage average is driven up by other sectors such as banking, engineering, and pharmaceuticals, while call center employees make much less. However, even if the average of $10,733 is used as a benchmark for a forty-five-hour week and a fifty-week year, it translates into about $4.79 an hour. Again—call center employees earn considerably less. Little wonder that U.S. airlines have embraced the subcontinent.

WNS travel clients include United Airlines and Travelocity, and there's no doubt how important this sector is to the company. "Travel is quite substantial," said Vyas. "It's our biggest vertical." He dismissed the "reverse outsourcing" trend highlighted by some U.S. companies that recently brought call center work back to America, and insisted the corporate clients are happy. But even if more airlines continue to do away with human interfaces and shift additional customer service to email models, Vyas said WNS will be ready: "So we have a solution for that, too. We provide all kinds of solutions."

The next day I was in Delhi, and this time I visited an even larger travel outsourcing office, belonging to the Bird Group, which also handles calls for Travelocity, as well as American Airlines. The firm's roots date all the way back to 1971, when it began fielding reservations as general sales agents for Lufthansa.

"The 1970s is when GSAs started taking shape," explains Ritu Bararia, the head of corporate communications.

Its site claims "Bird Group has created a niche in the market for specialized solutions for airlines, corporates, and travel agencies." The company maintains a workforce of more than 2,900 employees at forty locations throughout India, and generates more than $100 million in annual revenues.

Bararia noted that despite being "primarily a business-to-business company," Bird has expanded its own "verticals" even further than the standard outsourcing models, beyond call centers and IT to include airport ground handling, charter flight operations, and the hospitality industry. The firm owns a travel technology giant and operates its own training academy, which no doubt cuts down on U.S. executives flying in to instruct the staff. Bararia summed it up: "We're really big in the travel industry."

She and I discussed the changes we've seen in travel, and I asked her to explain why airline executives are smitten with outsourcing. "They look at us and they say, why not explore these options?" Bararia said. "Perhaps some of these options will work for us." And she sympathizes with her corporate clients: "Running an airline is a tough job."

But for many airline execs, companies such as Bird Group clearly make it less tough. Here, too, I was told that nearly all the employees are full-timers who receive health benefits, not transients who dabble at answering calls for a year or two before moving on to something else, as the PricewaterhouseCoopers report indicates. In fact, Bararia said, "We have a very steady workforce." And when I asked if U.S. travel companies are happy with their performance, she responded, "I am sure."

Outside the gate of one of these facilities, I tried to engage workers in conversation. Nearly all were friendly, though most demurred. But a young woman whom I will call Ms. Patel—simply because it's such a common surname in India—smiled when I explained why I was visiting. So, does she enjoy speaking to American travelers? Her smile grew: "Most are very nice. Only some are not as nice." What about her work conditions? She stopped smiling and moved her head in the direction of the ever-present clusters of young men who crowd the streets throughout India, often squatting, and apparently without much purpose. "I have a good job," she said, and then politely let me know she had to get back to work.

The simple fact is that I quickly came to like the people I met at the Mumbai and Delhi call centers. At one facility I was offered traditional Indian sweets because a staffer had just announced an engagement, and the receptionist at Bird Group who called for my taxi reminded me of Pam Beesly on *The Office*. The cubicles were outfitted with the same blond furniture from Ikea, the same Dell laptops, the same admins dressed in jeans with the same belt-clipped cell phones. The public relations people fielded the same calls about photo shoots and press releases. I noticed one desk held a coffee cup inscribed: TELL ME AGAIN HOW LUCKY I AM TO WORK HERE . . . I KEEP FORGETTING!

In other words, these are not aliens. Most Americans would like the workers I encountered. I met nice people, and they were clearly very proud of their nation's quick emergence into a technological giant. I came to believe that if some Americans are angry about U.S. jobs being shipped overseas, that anger should be directed toward the American executives who made those decisions, not at Ms. Patel or her coworkers.

The Cultural Divide That Dare Not Speak Its Name

While aircraft maintenance outsourcing is quite opaque to airline passengers, the outsourcing of call centers is something they encounter in a very transparent way. So there's another factor at play when our calls are routed to Southeast Asia, and it's an issue many Americans discuss only in whispers. A few years ago I tackled it head-on for USAToday.com: "When Customer Service Is Lost in Translation" dealt with the problems of hashing out travel snafus with reservations agents overseas. Talk about hitting a nerve; the responses were so overwhelming, and so filled with anecdotal evidence of a cultural divide, that I wound up writing a follow-up column as well.

The executives I spoke with in India didn't want to discuss it, but there's no getting around the cultural issues when a caller in Nebraska is routed to a customer service rep in Mumbai. That's not snobbery and it's certainly not racism; it's a simple realization that not all knowledge gaps can be bridged by a decision tree memo from an airline's marketing department. I found this out a few years ago when I called Delta and booked what turned out to be an invalid and expired fare offered by a rep in India who was monomaniacally obsessed with booking me a rental car I didn't need.

Often, the call centers make the problems much worse. With airlines increasingly closing their own call centers, many customers now need to rely on email communication, which poses several problems. Linda Burbank says, "A lot of people don't like email, especially older people. They're really frustrated because they don't hear back and have to wait many weeks." And a prompt response is a big part of complaint resolution. What's more, angry customers often cite the sensitive issue of deciphering the advice of foreign call center reps with heavy accents.

However, even U.S. airline executives have been forced to publicly acknowledge the obvious: many outsourced call centers simply haven't been up to the task. In January 2011, US Airways closed its call center in Manila, while Delta transferred work from an outside facility in Jamaica to its own offices in Atlanta and Minneapolis. One month later, United confirmed it was repatriating 165 such jobs from India to Chicago and Honolulu. Bloomberg quoted United spokeswoman Robin Urbanski as saying, "More sophisticated conversations with our guests are much better suited for us to handle instead of a third-party partner. We clearly have the deep industry expertise to help our guests navigate through their options."

No doubt many airline customers would agree. However, addressing the inherent and high-profile problems associated with call center outsourcing should not be viewed as a portent of a new attitude toward labor issues. U.S. airline executives have been more than happy to continue outsourcing many other functions. Clearly this trend has exacted a tremendous cost on America—economically and socially. But it may represent an even greater threat to the nation. In 2004 the National Intelligence Council warned of a "pervasive insecurity" because globalization will remake the planet, particularly the middle classes of the developed world.[1] For once the U.S. airline industry is cutting-edge. Because those prognostications have already come true for thousands of former airline employees.

When I reminded Secretary LaHood about President Obama's promise to stem the tide of foreign outsourcing, he responded, "This administration will do all that we can to maintain American jobs. Whether it's in the car industry or the airline industry or whatever form of transportation it is. We've encouraged people to come here and build high-speed trains. We've encouraged foreign countries to buy Boeing airplanes. Wherever we can encourage people to buy American and employ American workers, we do it."

No More Friendly Skies

For airline employees, the fun has long gone out of the business. Fun is harder to quantify than 401(k) annuities and dental coverage, but it's a very real factor nonetheless. While reviewing interview notes, it suddenly struck me how frequently veteran airline employees raise this issue on their own: working for an airline is just not fun anymore. "The harder we work, the more they say they don't need us," says Eadie Francis, a long-time employee of a major carrier.

And it's no coincidence the fun has gone out of the airline business for passengers as well. The airline service crisis is a direct reflection of the industry's labor crisis. Not all outsourcing takes place in India, China, or Mexico; every day passengers in U.S. airports interact with outsourced airline employees.

Outsourcing has become ubiquitous in the industry's bag rooms, since few carriers retain their own employees for such functions, and baggage handling is a lower priority even among outside service companies, what one expert calls "the hole no one wants to be in." Outsourcing only exacerbates the complacency, particularly if airline baggage handlers making $15 an hour are replaced with outsiders making $7 an hour. Therefore it's particularly telling that according to industry experts at SITA, 51 percent of all undelivered luggage is due to "transfer baggage mishandling," a service that is almost entirely outsourced at most major airports.

Other "employees" are right in passengers' faces. Etiquette expert Peggy Post, the former Pan Am flight attendant, cites another drawback of outsourcing: "For passengers, there is a problem—not with the people, but with their lack of training. They don't have that pride in the company." Unfortunately, this often manifests itself at the worst times, such as during widespread flight delays or cancellations, when it appears that no one wearing the airline's uniform is overseeing the crisis du jour.

What has become quite clear is that U.S. corporations, with the full and bipartisan blessings of Congress, the White House, and the Supreme Court, will continue to make decisions that are sensible to shareholders, not passengers, employees, or America itself. Glenn Tilton describes United's decision to delegate heavy maintenance on its Boeing 777s to Ameco in Beijing as "strategically very, very important." He also assails labor's efforts to prevent Boeing from shifting work on the 787 Dreamliner from a unionized factory in Washington state to a nonunion shop in South Carolina: "You cannot protect the status quo by litigating against change."

Where Did All the Jobs Go?

The numbers are not pretty. According to the DOT, in May 2011 there were 472,919 full-time employees in the U.S. airline industry; this represents 159,051 fewer full-timers than during May 2001, a 25 percent decrease. The Associated Press noticed the trend one year earlier, and reported the domestic airline industry "has now lost one of every four U.S. employees it had a decade ago." There's one glaring exception: Southwest Airlines has *never* involuntarily furloughed a single employee.

But even these numbers don't reflect the tens of thousands of layoffs and permanent furloughs airline employees have endured in the deregulated era. There have been three distinct shifts in airline employment in recent years. First, because of a workforce migration from bankrupt airlines to new-entrant carriers, many current employees are on their second, third, fourth, or fifth airlines, and have suffered multiple pay cuts along the way. Second, the growth sector in a stagnant industry has come from part-time employees, which have increased by nearly thirty-six thousand since 1990. Many positions barely offer minimum wage and virtually none offers health coverage, retirement, or meaningful benefits, sometimes not even nonrevenue flight privileges. And the final shift, of course, is from mainline airlines to regional airlines. Analyst Bob Mann calls it "switching inexperience for experience" and has tabulated the results of what he terms "job transferency." In 1999, about 95 percent of U.S. commercial pilots worked for majors and 5 percent for regionals; by 2008 those numbers had become 70 percent and 30 percent—and the gap keeps widening. As for the staggering pay differential, chapter 6 details how the Colgan Air first officer in the 2009 Buffalo crash earned $15,800 annually and had moonlighted as a waitress.

Little wonder that in April 2011 *U.S. News & World Report* disclosed "10 Fields Where Workers Are Falling Behind," and airlines ranked as having the largest decline; the average hourly pay of $26.55 represented a 4.2 percent decrease since 2007. And the hits just keep on coming. In cities across America, economies have tanked as airlines have downsized and outsourced.

In January 2011, the *Pittsburgh Post-Gazette* documented US Airways cutting another ninety jobs in that city; at its peak the carrier employed 12,000 people in the Pittsburgh region and by last year that number had fallen to 1,883. The head of a local International Association of Machinists chapter was quoted as saying he suspects the work "will be subcontracted out to people making lower wages." He also stated the union was "powerless to do anything about the layoffs."

Meanwhile, the compensation enjoyed by senior airline executives continues to grow, regardless of whether the company is bleeding red ink or the rank and file are enduring pay cuts. According to UpTake Networks, in 2009 the CEOs of the ten largest domestic carriers earned $34.1 million while their companies collectively lost $3.3 billion. Of course, this is part of a larger trend: the top S&P 500 CEOs make about 344 times what the average American worker makes, an astronomical jump from the 40-to-1 ratio of 1980. But it's critical to note that unlike nearly all domestic airlines, most of those other corporations have been profitable over time. The argument that large paychecks are necessary to attract executive talent rings hollow at carriers that have not been profitable for years on end.

The two airline labor issues—dwindling union membership and skyrocketing CEO compensation—are not unrelated. As the number of airline employees in unions has fallen, CEO pay has climbed. In fact, between 1980 and 2005 the number of unionized workers was slashed from about 20 million to 7.5 million, while those CEO salaries soared.

Even the financial meltdown of 2008 didn't alter the course of events: the *Wall Street Journal* reported bonuses for CEOs at fifty major corporations in 2010 increased by 30.5 percent over the previous year. And as for the airlines, if you examine the deals given to senior executives, you'd never know it's an industry hemorrhaging money. According to the AFL-CIO's CEO Pay Database, the total salaries for those running the nation's ten largest public airlines comes to $38,907,562. Conversely, the AFL-CIO reports the median salary earned by U.S. airline workers is $33,190. So, here's how the CEO's pay at each of those same ten airlines compares with the average employee's W-2:

Delta/Richard Anderson	252 times more
Hawaiian/Mark Dunkerly	191 times more
American/Gerard Arpey	179 times more
United Continental/Jeffery Smisek	131 times more
Southwest/Gary Kelly	101 times more
Alaska/William Ayer	101 times more
US Airways/W. Douglas Parker	77 times more
AirTran/Robert Fornaro	60 times more
Republic/Bryan Bedford	40 times more
JetBlue/David Barger	36 times more

As the Massachusetts Institute of Technology's William Swelbar notes, salary often represents about 20 percent of a CEO's compensation and must be considered along with bonuses, stock awards, preferred stock options, payments to savings plans, and other rewards. He suggests that CEO compensation is "volatile" because it is risk-based and dependent on market conditions; he further argues that compared to American industry at large, airline employees rank fairly well: "On balance, jobs in the airline industry pay quite well as they relate to jobs that require similar training and experience across the employment spectrum."

It's an argument I can't accept. For one thing, U.S. airline executives are overly compensated regardless of whether their companies perform well or not. For another, I know what it's like to load baggage into the belly of a hot airplane on a sweltering summer day, and no executive decision-making is worth 252 times that compensation.

Contentious Relations in the Air

In the aftermath of deregulation, the dramatic increase in commercial flights led to stress on the system itself, with 337,000 more departures between 1978 and 1980; the increased workload manifested itself in record cases of hypertension and stress-related ailments among members of the Professional Air Traffic Controllers Organization. When controllers went out on strike in the first year of the Reagan administration, the new president promptly fired nearly all PATCO members and the union shortly dissolved. (Like many airline veterans, I refer to "Ronald Reagan Washington National Airport" as "Washington National" on the grounds that [a] one U.S. president is enough for any title and [b] anyone who fired the nation's air traffic controllers should not have an aeronautical facility named in their honor.)

It's worth noting that the immolation of PATCO had far-reaching effects, beyond aviation and even beyond government employment to the larger free market itself. Alan Greenspan, the ostensibly neutral former chairman of the Federal Reserve Board, gushed about PATCO at the Ronald Reagan Library in 2003:

> But perhaps the most important, and then highly controversial, domestic initiative was the firing of the air traffic controllers in August 1981. . . . There was great consternation among those who feared that an increased ability to lay off workers would raise the level of unemployment and amplify the sense of job insecurity. It turned out that with greater freedom to fire, the risks of hiring declined. . . . Whether the average level of job insecurity has risen is difficult to judge, but, if so, some offset to that concern should come from a diminished long-term average unemployment rate.

Difficult to judge? The effects of Corporate America's "greater freedom to fire" are not at all difficult to judge for anyone in a different socioeconomic class from Alan Greenspan. Particularly since if in fact there has been a "diminished long-term average unemployment rate," it has been generated in large measure through lower-paying jobs.

Make no mistake, there truly is a war on organized labor, and it's certainly playing out in Washington. Consider that in June 2011, Congressman John L. Mica of Florida, the Republican chairman of the House Transportation and Infrastructure Committee, released a snarky statement—"Big Labor Captures Airport Screeners"—that asserted: "The traveling public will be absolutely delighted to learn that big labor has captured the TSA's army of airport screeners. . . . The Obama Administration today can check the box on another boost for big labor and a significant setback for the traveling public." Mica's staff did not respond to requests for an interview.

If battle lines are being drawn, I move from one side of the skirmish to the other, interviewing airline executives and financial analysts, as well as labor officials and academics. But in the annals of the American labor movement, one name undoubtedly stands out from all others: Hoffa. So I traveled to Washington.

Once inside the office of James P. Hoffa, the general president of the International Brotherhood of Teamsters, I was equally struck by the spectacular picture-window view of the Capitol dome and the striking resemblance to his father, James R. Hoffa. As he showed me the display of model airplanes representing the union's airline contracts, I noticed that his desk held a stack of DVDs featuring Jack Nicholson in the eponymously titled biopic of his father's life.

These are challenging days for all union leaders, regardless of their familial pedigree, even though Hoffa boasts of representing workgroups ranging "from airline workers to zookeepers." There are others within the Teamsters who are focused solely on airline issues, and I spoke to them as well. But I asked to visit Hoffa because I wanted a big-picture perspective, and his opening remarks were as expansive as our view of the Washington Mall. He spoke of the American Dream and Corporate America's pursuit of the almighty dollar and the effects of deregulation on the trucking industry (the number of Teamsters truckers has shrunk from about 400,000 in 1979 to about 60,000 now, due to both deregulation and free trade agreements that allow truck drivers to transit "from Mexico to fucking Montreal"). His vitriol was expressed by referencing "NAFTA and CAFTA and SHAFTA" and pending bills in Congress to expand free trade to even more countries—Panama, Colombia, South Korea. And he easily rattled off the names of dozens of companies that have shuttered American factories and uprooted to Latin America or

Asia: Oral-B and Mr. Coffee and Hershey's and even Swingline Staplers from my home borough of Queens.

"Corporate America has betrayed America and American workers," Hoffa said. "Executives don't see their obligations. All they think about is the next quarter. And they have a date when they're going to leave. They don't think beyond that. They feel they don't owe one thing to this country."

We settled in to discussing aviation and I told him what both current and former airline executives had been telling me: the airline industry's labor problems are due almost exclusively to union greed. "There's a way to pay fair wages," responded Hoffa. "A 747 pilot flying to Japan with three hundred people onboard? Look how productive he is. He should be well paid." As for work rules, he maintained that the Teamsters and other labor organizations are certainly ready to negotiate on work rules.

Hoffa has been battling with corporate management for decades, but clearly he and all other union representatives know there's been a radical shift in the 2010s—and it's not hyperbole to suggest it may be nothing less than "a war on workers." Hoffa said, "The airlines are part of a bigger picture. They're moving back all borders. And American corporations are dodging taxes. That's why there are no jobs. And that's why there is no money. Why are we closing libraries? There are no tax dollars." He further suggested that even if the U.S. government agreed to radically reduce corporate taxes from 35 percent to 10 percent, American companies would continue to outsource to developing nations.

A running "joke" at the FAAC meetings was that United chairman Glenn Tilton would grow apoplectic at any mention of increasing the industry's tax burden. But of course the simple truth is that the airlines—like most U.S. corporations—have seen that burden dramatically reduced in recent decades. When the *New York Times* reported in March 2011 that General Electric paid *no* U.S. taxes in 2010—as in zero dollars, zero cents—it also revealed that GE had kept its foreign profits offshore. What's more, the Urban Institute and Brookings Institution recently noted that revenue from corporate taxes fell from between 5 percent and 6 percent of GDP in the early 1950s to 2.1 percent of GDP in 2008.

Gordon Bethune, the former CEO of Continental who is widely credited for turning that airline around, openly acknowledges such tax policies are the reason American companies build, hire, and invest overseas; this includes the aerospace giant Honeywell, where he sits on the board. He bluntly warns that President Obama's pleas that corporate executives hire more Americans are falling on deaf ears, and suggests instead the loss of U.S. jobs can be solved by adjusting tax laws to make the U.S. more attractive for investment by corporations: "It's not a welfare state. Corporations don't like hiring people they don't need."

No doubt. But how much more attractive? In fact, we have to wonder if this is yet another example of corporate depravity enabled by companies given the rights but not the responsibilities of citizens. An individual who lives in America but makes a fortune by hiring cheap labor in China and then in turn invests all his profits elsewhere would be considered unpatriotic, a tax dodger, a weasel. But an American corporation that engages in identical behavior is considered smartly run.

I posed this question to Glenn Tilton of United/Continental: Doesn't it make sense—from a business rather than a patriotic perspective—to help ensure that Americans have jobs, since those same Americans are the passengers most likely to be flying on United? He pointed out that his company's share of the domestic market is about 25 percent, and added, "There's more demand for our product outside the United States. And the economies are growing faster outside the United States." He said the "dilemma" stems from America's stagnant and mature economy.

The View from the Top: Management Blues

Of course, the FAAC needed an airline financial analyst and that fell to Dan McKenzie. Six months after the FAAC adjourned, I met with McKenzie in his office overlooking the Chicago skyline, just a few blocks from the world headquarters of United/Continental. He stressed that he has nothing but love and respect for labor leaders. But he asserted that employee groups put too much emphasis on seniority: "I expect my one-year captain to have the same ability as a twelve-year captain. The difference between a one-year surgeon and a twelve-year surgeon is that the twelve-year surgeon is more productive and can perform surgery in three hours instead of four hours. But that's not the case with pilots. It's a fatal legacy issue—they need to get this."

McKenzie and I launched into a rambling conversation about work and compensation and the nature of fairness. I posited that an employee at any company who remains loyal over time and works hard and plays by the rules deserves to be rewarded periodically, no? Aren't raises the American way? "That is not the American way," McKenzie replied. "The American way is a meritocracy." While he does recognize the need for pay to keep pace with inflation, he added, "I don't believe in job protection. And I've lost my job twice, as painful as it was."

However, even Wall Street recognizes that the labor well is running dry; analyst Helane Becker acknowledges, "Our view is the airlines have probably gotten as much as they can get from their employees." She notes that one workforce group at American Airlines was earning 1998 wages in 2011, and adds, "You can't continue to beat up your employees." And there's another factor at play. Low-cost airlines—particularly those that started up in the last decade—have an inherent labor advantage since they're not dealing with a workforce that has been in existence since, say, 1924 (Delta) or 1926 (United) or 1929 (American). However, the challenge for low-cost carriers such as Frontier (1994) and JetBlue (2000) and Virgin America (2007) is that as they mature, their employees mature as well; that's why Becker points out, "You can't fire your people every five years."

One management solution was devised when Bob Crandall headed up American Airlines: keep existing pay scales in place, but pay newer employees at a lower level. At the risk of being tossed out of his Florida winter home, I asked Crandall point-blank: You're the guy credited with introducing the B-scale, so aren't you responsible for what's happened with airline outsourcing? He replied: "No. All of the people in that second tier would eventually get, not to the high point, but well into the first tier. And they were never hired at the sort of outsourcing rates that people are hired today. They were simply hired at the same rates as Southwest and the start-up airlines were paying at the time. The whole rationale was we're going to create a new airline and be competitive with the new guys, but we'll keep the old airline and the new airline. And we'll protect the old airline people because the new airline people will allow us to lower our costs." One rank-and-file American employee with nearly thirty years of experience agrees: "Crandall was a son of a gun, but he was our son of a gun. He did close the gap with two-tier wages. It took a while, but he kept his word."

The industry's notorious work slowdowns also fuel management anger. In the spring of 2000, United trailed all other domestic carriers with a 57 percent on-time arrival rate, which was due to both pilots and mechanics refusing to work overtime after their contracts expired. (We experienced a similar situation at the Pan Am Shuttle with unhappy pilots, who one day began declaring all departures scheduled for thirty minutes past the hour as having pushed at thirty-six minutes past, knowing that internally anything more than five minutes was a delay. Our solution in

Operations was to simply "mishear" their departure calls and record them all as thirty-five minutes past, which did not earn us much love from the pilots.)

But it's not exclusively about pay. Swelbar may be speaking for all airline boards when he says that labor contracts and work rules have caused the industry to employ more people than it has needed, thereby creating higher cost structures.

Current airline executives won't speak at length about their labor woes. But former execs will, particularly when it comes to the most expensive workgroup—pilots. Michael Levine says pilot unions are rabidly fixated on two topics: allowing their members to bid for their monthly flying time by seniority and tying pay to the size and type of aircraft flown, so more experienced crews fly the largest jets. Levine says, "These are things that the pilots union fought for in 1934, and they're not going to let go of. There's no reason why you should pay a pilot by how many seats are strapped to his butt. There's no reason why a senior pilot shouldn't be able to fly the shuttle and make whatever money he's supposed to make."

Bob Crandall and others believe Southwest's unions have behaved differently, by increasing their efficiency in exchange for more compensation. Then why, I asked, don't other airlines emulate Southwest's policies? "The [other] unions won't take it," said Crandall. "The unions at the [legacy] airlines were all created after World War II. And they got used to a certain set of conditions and they wouldn't give up those conditions. But they look at how much Southwest is making and they say, that's how much we want to make but we don't want to do that much work."

Another former executive who is quite blunt about labor relations—even going so far as referring to some union representatives as "Sky Nazis"—is former Continental CEO Gordon Bethune. He maintains that his contentious relationship with the

Association of Flight Attendants was due to inflexibility on the union's part, in not being amenable to rule changes that would help the airline. He also says, "Airlines with the worst consumer records have the worst labor records."

A retired flight attendant told me that her old colleagues are all unhappy—but she warned that you need to sift through what is legitimate and what is whining. In fact, I've spoken to former airline employees who are as bitter toward their former unions as the airline executives. One is R. Bruce Silverthorn, who spent decades in the airlines as a ramp worker and customer service agent, beginning at Mohawk Airlines in 1963 and continuing with Eastern and Delta. His diverse experience on the ramp and in customer service—as both a union and nonunion employee—gives him a unique perspective on management-labor struggles. I met him for beers at a bar in Buffalo, and soon realized it would probably be best to slightly edit his more colorful quotes (Silverthorn cannot resist using ten-letter and twelve-letter descriptions when referencing Frank Lorenzo, the notorious corporate raider who ran Eastern into the ground).

"I don't dislike unions," said Silverthorn. "But I think Eastern's unions fucked up. They sold us down the river. . . . The unions are as much to blame as Lorenzo." In his view, union greed helped sink the company, even as labor officials protected "deadwood" employees who did not produce. Even so, he believes outsourcing has made things worse for customers: "Before outsourcing there was a pride. Now they don't give a shit."

The View from Down Under

When I spoke to Ralph Nader about the dramatic influx in maintenance outsourcing, the first thing he asked was, "Why are the unions allowing this?" Perhaps because they can't stop it. They've expended quite a bit of energy in recent decades fighting airline executives, bankruptcy judges, and even Congress, and in some ways the big unions are exhausted and even spent.

As one mechanic said to me about airline management: "They're so worried about the shareholders. But it's like a marriage—you have to take care of your employees. If you don't, someone else will. If you treated your people right, there wouldn't be a need for unions."

Battling deep-pocketed airline CEOs has taken its toll. Among the rank and file, for example, it's hard to underestimate the industry's vitriol for Frank Lorenzo, the former head of Texas International who founded New York Air and also acquired Frontier and PeoplExpress. Even encapsulating his reign isn't easy: he also took over both Continental and Eastern, drove both major carriers into bankruptcy, oversaw one of the industry's most bitter strikes, and pled guilty to safety violations at Eastern. In 1989, Barbara Walters dubbed him "probably the most hated man in America." And even William F. Buckley commented: "As a businessman, Frank Lorenzo gives capitalism a bad name."

When Lorenzo attempted to reenter the business in 1993 by launching a new company with the possibly tongue-in-cheek moniker of "Friendship Airlines," labor unions and even Congress howled. Former FAA administrator Randy Babbitt, then president of the Air Line Pilots Association, stated: "Too much is already known about his questionable business practices to allow him to avoid disclosure in this case." The DOT took the extraordinary step of denying Lorenzo's bid, citing the FAA's repeated involvement in resolving safety questions at Eastern from 1987 through 1990.

Just coming up short to Lorenzo on the judges' cards in "Most Hated Airline Executive" polling is corporate raider Carl Icahn, one of the famed 1980s barbarians at the gate and the engineer of an infamous hostile takeover of TWA. Despite concessions from

TWA's unions, Icahn apparently sought further givebacks while downsizing the airline and selling off routes.

By 2001 it all came apart and TWA entered bankruptcy for the third time, with its assets sold to American. Employees considered the sale bittersweet because although many now had jobs at American, pre-merger plans for expansion of TWA's maintenance base in Kansas City never materialized, and the seniority integration meant thirty-year TWA veterans were being treated as new hires. TWA's labor woes live on long after the airline folded, with the pilots unions still feuding.

Pat Friend, the longtime president of the Association of Flight Attendants, maintains there's a strong link between the string of airline bankruptcies and labor relations. Thanks to bankruptcy judges, management got a taste of not having to bargain, and now many companies don't want to come to the bargaining table anymore.[2] Friend also maintains that the National Mediation Board members are not doing their jobs; six months after flight attendants at Delta-Northwest rejected the AFA, the union filed a motion but the mediation board still had not responded.

Friend asserts that the real changes in the industry took place in the late 1980s and early 1990s, when innovations such as self-ticketing kiosks and checking in for flights at home took hold. Many industry insiders believe such programs were always about the reduction of staff, not the convenience of passengers.[3]

On the flip side, for union leaders it's about employees making a decent living. Today Robert Roach is the general vice president of transportation for the International Association of Machinists, but he began his airline career on the ramp at JFK in New York for TWA. With all he has seen both before and after deregulation, I asked him if labor conditions will only worsen in the long term. "Yes," he replied. "Unless something changes. Workers want to make a decent wage and go home and focus on their families. But [executives] are cannibalizing the industry."

Meanwhile, Roach said no Northwest employees were laid off when the airline was acquired by Delta. But I pointed out that Northwest employees in the IAM were laid off *prior* to their company's merger with Delta; in smaller "spoke" cities longtime employees were given two weeks' notice and asked to train their outsourced replacements! A former Northwest employee told me: "They voted all the spoke cities out of jobs. Including one guy who was three months from turning fifty-five and retiring." Northwest later issued its infamous "garbage memo" to laid-off employees, recommending they rummage through used clothing bins and Dumpsters to make ends meet.

In addition to downsizing and outsourcing, Roach points out that Delta's "new thing" is being carefully watched throughout the industry. The airline's customer service positions are increasingly being filled with "ready reserve" employees who are severely restricted in the number of hours they work and receive neither health coverage nor pensions and only "limited travel privileges." They perform a variety of cross-training functions, so the requirements include driving company vehicles, typing, and "lifting up to 70 lbs. repetitiously." According to Delta's site, the pay is $9.07 nationwide (and slightly higher at three large airports), but the ready reserve force is not on track for promotions. Basically they make themselves available on certain days and then—like substitute teachers—wait for a call that may or may not come.

However, at least those workers are employees of Delta. Increasingly, the airline industry turns to outsourced firms to fill frontline customer service jobs. As Roach explains, "They have people dress up in company uniforms. I've spoken to workers who don't know what they're doing."

I asked Roach which domestic airlines, if any, enjoy the best reputation among labor leaders, and he immediately responded with Southwest and Continental. However, Continental is being melded into United, so Roach acknowledged that "the jury is still out" on whether good relations with CEO Jeffery Smisek will continue at the newly merged corporation.

As for Southwest, it's widely recognized that the company maintains the best labor relations among all U.S. airlines. (It also is very heavily unionized, a fact many observers find surprising considering Southwest's low-cost image.) Besides being the only consistently profitable carrier in America for the last three decades, Southwest ranked twelfth and was the only domestic airline to crack the top fifty in *Fortune* magazine's 2010 list of the World's Most Admired Companies (Singapore Airlines ranked twenty-seventh).

On Not Passing the Baton

When I worked for Tower Air, we used to joke that someday the FAA would allow airlines to outsource dispatchers. Ha. And mechanics. Ha. And flight attendants. Ha! And pilots! Ha ha! To varying degrees, all that is occurring now.

What we didn't foresee is that massive layoffs and massive outsourcing would affect not only labor but customers as well. And it certainly affects the quality of the products. Without that mentor-protégé relationship, how will the next generation evolve? The piloting heroics displayed by Captain Al Haynes in Sioux City and Captain Chesley Sullenberger on the Hudson River were developed over long and meticulous career paths. As Haynes says of United Flight 232, "We had 103 years of experience [in the cockpit]. If we couldn't do it, then it couldn't be done." He also notes advanced technologies present new problems: "Now I'm concerned about the effects of the [computerized] glass cockpits. Pilots are not learning to fly by the seat of their pants. So if something happens to the systems, they don't know what to do."

David Bourne, director of the Airline Division for the Teamsters, concurs. He began as a regional carrier pilot himself, a twenty-three-year-old flying Beech 99s for Allegheny Airlines and making $12,000 a year—back when that was enough to live on. The minimum number of flight hours—a primary hiring qualification for commercial pilots—was much higher in past decades, and Bourne did everything including towing banners along beaches to accrue more cockpit time. He cites regional captains who have barely more hours than their first officers, and notes, "Now you have nobody to learn from. We've lost a lot of that. We don't have that osmosis."

Perhaps the cruelest manifestation of this syndrome is the military's recruitment efforts, which promise bright futures in the civilian sector. For generations, young Americans have entered the service and been trained in high-tech aviation skills, knowing the risks and hardships were mitigated somewhat by the security of a bright airline career ahead. But as Bourne points out, returning veterans from Iraq and Afghanistan are learning a bitter lesson: the best U.S. airline jobs are being offshored from the country they fought to defend.

6

When Your Airline Isn't Your Airline: Regional Carriers Provide Lower Levels of Service and Safety

> All airlines providing scheduled service in aircraft with 10 or more passenger seats operate under identical FAA regulatory requirements.
>
> —*Regional Airline Association*

AT ONE PARTICULARLY CONTENTIOUS MEETING of the DOT's Future of Aviation Advisory Committee in November 2010 I attempted to convince a room filled with airline industry executives and DOT officials that the widely accepted industry practice of outsourcing flying to regional airlines is, on its face, duplicitous.

My point was that most consumers have no idea that when they buy a ticket on Airline A, they are likely to be flying on Airline B. That term *codesharing* is most prevalent in the form of regional operations, which is a complex way of stating that the nation's largest airlines have elected to outsource and subcontract much—and sometimes even most—of their flying. At that meeting in Washington I argued that if we stopped one hundred passersby on street corners in any town in America, only a handful would understand the term *codesharing.* I argued that most passengers don't realize they are flying on regional airlines, despite the incredible fact that such flights now represent 53 percent of all airline departures in the United States every day. There simply isn't enough transparency in the system, because most consumers don't understand that a booking made through Continental or United will generate a ticket for a flight operated by Colgan Air.

Finally I decided to show and tell. I reached into my bag and confessed to stealing a safety-critical item off a commercial flight that morning: the "ditching card" containing the emergency evacuation instructions found in the seatback pocket. I explained how I had booked my flight on Delta.com; that my Visa card was debited by Delta; that Delta sent me email confirmations with my flight details; that the taxi dropped me off at the Delta terminal, where I checked in at the Delta counter under Delta signage, and spoke to an employee wearing a Delta uniform, and was issued a boarding card from a Delta kiosk; I double-checked my Delta departure on the flight information display screen, and proceeded to a Delta gate decked out in more Delta signage; I spoke to yet another employee wearing a Delta uniform and boarded through a Delta jet bridge; I stepped onto an aircraft painted in the colors and logo of Delta and passed a flight attendant dressed in Delta's corporate colors; finally I took my seat and retrieved the ditching card, which prominently displayed Delta's red-and-blue "widget" logo on the cover. On the back page, in considerably smaller font, were the words: "Operated by Atlantic Southeast Airlines."

Some of my FAAC colleagues and the many DOT staffers in the room seemed a bit puzzled. After all, what passenger has never heard of Atlantic Southeast? Or Chautauqua and Mesaba and PSA and Air Wisconsin? And who has never heard of codesharing? And who doesn't know that the FAA polices, inspects, and fines airlines based on the companies holding the operating certificates, not the names painted on the airplanes? And who doesn't know that outsourced regional airlines in turn outsource aircraft maintenance and other key functions? Who doesn't understand how the airline industry works these days?

I begged to differ. Yes, there have been positive steps taken in recent years to educate consumers about codesharing and regional flying. But I argued that in far too many cases, passengers and their families become aware of such corporate partnerships only after a regional airline operating on behalf of a major airline is involved in a fatal crash.

Two Sets of Rules: Lower Standards = 50 Deaths

Unlike many next of kin on September 11, 2001, Beverly Eckert got to say goodbye. Sean Rooney was at work on the 105th floor in the South Tower of the World Trade Center that morning and both he and his wife—watching events unfold on television—knew he would not make it out alive, so they spoke to each other of love and loss right up until the moment the building collapsed. Almost immediately, Beverly threw herself into a new cause, advocating for enhanced cockpit security and closing other gaps exposed by the terrorist attacks. She became immersed in the 9/11 Family Steering Committee, which was influential in forming the 9/11 Commission. Seven and a half years later, on February 12, 2009, Beverly Eckert would also perish in the heat and light of burning jet fuel.

That evening Continental Connection Flight 3407 departed Newark in icing conditions for a fifty-three-minute flight to Buffalo, but crashed about five miles short of the Buffalo airport into a home in Clarence Center, New York. Beverly and the forty-eight others on board were killed, as was a resident of the house hit by the Bombardier DHC-8-400. Beverly was one of five children, and in the immediate aftermath of her death two of her sisters retired from their government jobs and emulated her by becoming activists—not for preventing another 9/11, but for enhancing the safety of U.S. regional airlines.

"We're not professional victims," explains Karen Eckert. "We're advocates who want change." She and her sister Susan Eckert Bourque have joined with others who lost loved ones on Flight 3407 in lobbying for stricter regulation of the airline industry's increasingly darker side: the regional aircraft operations that now make up 53 percent of all commercial airline departures in the United States. That's because those fifty people were killed not in a Continental Airlines crash, but rather in the crash of a flight operated by Colgan Air, a company many people had never heard of before 2009, including some of the passengers on the fatal flight.

Ultimately the National Transportation Safety Board stated: "The probable cause of this accident was the captain's inappropriate response to the activation of the stick shaker, which led to an aerodynamic stall from which the airplane did not recover." In the immediate aftermath of the crash, the entire nation would learn quite a bit more about the cockpit crew of Flight 3407, about Colgan Air, and about the regional airline industry itself. The reports emerged in spurts, but eventually an image coalesced. The crew's fatigue and the long days they spent commuting; the napping in crew lounges and crash pads; the first officer's commute from her home in Seattle to her crew base in Newark before her flying day even began. Still more: the captain's record of four check flights failed by senior management and unsatisfactory proficiency checks. The cockpit voice recorder transcript indicating the first officer earned a gross salary of $15,800 in 2008 (necessitating that she moonlight at a coffee shop). The "unsterile cockpit" environment that included text messaging and personal conversations during critical phases of flight. Later, the NTSB would report that when hired by Colgan, the captain had accumulated just 618 total flight hours and the first officer 1,470 total flight hours.

"We are toying with disaster," warns FAA whistle-blower Ed Jeszka, a former flight instructor. "I've never heard of a pilot not learning how to deal with a stall." What's more, he points out that the regulations passed *after* the Buffalo crash would not have prevented that tragedy.

Throughout the industry, professionals were stunned. Why would an airline captain respond in the worst possible way to a stick shaker, a standard aeronautical device attached to an airplane's control yoke and designed to warn a pilot of an in-flight stall? But the shock soon spread beyond, to legislators and journalists and even passengers. Where were the industry's hiring standards for pilots with so little experience? Where was the training for icing conditions and stall recovery? Where was Colgan Air in all of this? Where was Continental? Where was the FAA? And not for nothing—could a commercial airline pilot in the United States in the twenty-first century truly be earning less than twenty grand a year?

The Eckerts are critical to my own understanding of what we as a nation will—or will not—learn from Flight 3407. We both have siblings working in government service, so I reached out to the Eckert sisters through this connection, not the advocacy group to which they are devoted. One holiday weekend I spent nearly four hours in Karen's dining room, and our meeting began with all three of us articulating the pain of losing a brother or sister. But no tears were shed.

We talked quite a bit about Beverly, and a strong image emerged. Clearly she was both passionate and compassionate, forceful when fighting for a cause, and tireless in her devotion. I asked Karen if Beverly was aware her final flight was operated by Colgan Air, and the response was a strong no. Because she lived in Connecticut and often visited family in Buffalo, Beverly was a frequent flyer who usually opted for nonstop JetBlue flights from JFK. But that evening she was using an expiring voucher on Continental; when her sister described the local weather as icy, Beverly replied, "Ice? Oh no!" Karen recalled, "I don't think Beverly realized she was not going to be on a jet. It probably gave her pause."

Even after family members were notified of the crash, Susan remembers being told by a Continental employee to wait for a team from a company she had never heard of: Cougar? Culligan? Only then did most of the relatives learn that Flight 3407 was operated by Colgan Air, which the families came to refer to as "a farm team." They felt the codesharing was a bait-and-switch.

They are not alone. In fact, it's a sentiment that has been repeated time and again during NTSB hearings. Meanwhile, the DOT and FAA repeatedly state there has not been a fatal "major" airline accident since 2001, when American Airlines Flight 587 crashed in the Rockaway section of Queens, in New York City. But consider that since that time we have witnessed six fatal accidents of smaller carriers:

- 2003: Air Midwest Flight 5481 in Charlotte, North Carolina; 21 dead; operating as US Airways Express
- 2004: Pinnacle Airlines Flight 3701 in Jefferson City, Missouri; 2 dead; operating as Northwest Airlink
- 2004: Corporate Airlines Flight 5966 in Kirksville, Missouri; 13 dead; operating as AmericanConnection
- 2005: Flying Boat Inc. Flight 101 in Miami; 20 dead; operating as Chalk's Ocean Airways
- 2006: Comair Flight 5191 in Lexington, Kentucky; 49 dead; operating as Delta Connection
- 2009: Colgan Air Flight 3407 near Buffalo; 50 dead; operating as Continental Connection

It's important to note that from the FAA's standpoint, none of these six aircraft was operated by a "major" airline, a carrier defined by the DOT as posting annual revenues of $1 billion or more. However, five of these six crashes involved regional carriers operating on behalf of five different major airlines—US Airways, Northwest, American, Delta, and Continental. In each of these cases, passengers bought their tickets through the major carriers, not the regionals.

The day after I met the Eckerts my sister took me to a memorial plaque in Clarence Center, and then to the crash site itself. I was surprised to see that such a small plot of land on a quiet residential street actually entombed fifty people. After I returned home, Karen sent me an email saying: "The loss of these precious human beings is immeasurable, both to us—their families and loved ones—and to the lost potential of their lives to make a positive difference in our country and our world. These were not expendable people who had knowingly, willingly put themselves at risk. They had no way consented to boarding a plane with fatigued, poorly trained, poorly paid pilots who lacked a disciplined safety culture in the cockpit, on an airline whose name is not reflected in the name 'Continental Flight 3407' or the Continental colors painted on the tail."

The Minor Leagues: Major Airlines Outsource Flying

When I worked in the industry we used to call them puddle jumpers, weed whackers, even lawn mowers. Pan Am Express flew funny-looking ATR 42s that were widely referred to as enchiladas. The implication was these small, slow, noisy prop airplanes were second string—not as comfortable, not as fast, and statistically not even as safe. Today many of those turboprop aircraft have been supplanted by regional jets, commonly known as RJs; overall, regional aircraft have gotten larger in recent years. Turboprops now carry between 9 and 78 passengers, while regional jets are configured for 30 to 108 seats. But serious concerns about comfort, reliability, and even safety and security still remain.

The terms *regional* and *commuter* and *feeder* occasionally are used interchangeably within the industry, and now some within the industry are calling them *network extenders*. But by any name, their presence is growing: U.S. regionals carried 159 million passengers in 2010, almost double the 82 million passengers carried in 2000. And regionals provide the *only* scheduled airline service at some 484 airports nationwide.

A regional operation traditionally referred to a major hub-and-spoke airline that subcontracted to a smaller airline to provide "feeder service" by flying "thin" routes, usually to less populated or rural markets. But in recent years the routes have gotten thicker; in fact, at certain times of the day six out of seven flights between New York City and Washington, D.C. (not exactly Lubbock and Odessa) are now operated by regional aircraft. As major airlines focus on more profitable long-haul and international routes, they're outsourcing more and more domestic flying to regionals. As analyst Bob Mann puts it, "There's a lot of interest in flying people from Moline to Abu Dhabi. But there's no interest in flying them from Moline to Chicago."

No doubt much of the confusion and misinformation over regional airlines stems from the difficulty in creating an up-to-date and accurate scorecard of who is flying for whom. The challenge is how best to illustrate the complex relationship between the nation's mainline airlines and the regional partners to whom they outsource so much flying. Excel chart? Venn diagram? Foldout schematic?

For 2012, the current dance card for the airline industry entails seven large airlines outsourcing flying to twenty-two regional carriers (prior to the melding of the recently merged United and Continental; all the details are provided in Appendix A). But it's worth noting that Delta and United partner with eight different regional airlines each, while US Airways partners with nine regionals.

Confused yet? Little wonder that William Swelbar, the MIT research engineer who analyzes the aviation industry, maintains that the next wave of consolidation in the airline business will take place among the regionals: "I believe we're at a point where there are too many." So to further muddle matters, imagine how confusing it is for the regional airlines themselves, since in many cases they are operating for companies that compete with each other, and sometimes they compete with themselves. Chautauqua, to cite just one example, partners with five mainline carriers: American, Continental, Delta, US Airways, and sister company Frontier. (A complete breakdown is provided in Appendix B.)

Of course, all this is subject to change, since these partnerships are struck—and struck down—on a constant basis. In some cases, regionals are owned by major airlines and/or their parent companies. US Airways Group owns Piedmont and PSA, while Alaska Air Group owns Horizon Air. And AMR, American's parent, also owns American Eagle but not its AmericanConnection partner, Chautauqua. On the other hand, quite a few regional carriers are sister companies; Pinnacle Airlines Corp. owns Colgan, Mesaba, and Pinnacle. And Republic Airways Holdings, which owns Frontier, also owns Chautauqua, Republic Airlines, and Shuttle America. I wish I were clever enough to have fabricated this, but it's all true.

What's become clear is that even the most frequent of flyers do not understand all these complexities. And make no mistake: there are real and tangible differences between mainline airlines and regional airlines, and real and tangible concerns over their growing presence in American aviation. These concerns include sales transparency issues that mask duplicity to consumers; service and performance standards that imply a bait-and-switch; and onboard comfort and amenities that are not up to par. What's more, gates and airports and skies are being clogged by smaller airplanes, and those planes are producing more carbon emissions. Finally, industry experts note there are two levels of safety among mainlines and regionals, not one as the FAA maintains.

Transparency: What Airline Did I Just Buy a Ticket On?

Commuter carriers certainly have their detractors, but they could not have a stronger advocate than Roger Cohen, president of the Regional Airline Association. In fact, I found myself laughing at the skill he displays in finessing every widely perceived drawback into an advantage. Here are some highlights from my lengthy conversation with him comparing regional airlines to their major airline partners:

- Don't regionals offer fewer onboard amenities? "Regionals and mainline carriers have morphed together now that there are no meals."
- Aren't the cabins smaller? "That provides easier access. And there are no middle seats."
- But the overhead bins are smaller, too? "Checking a bag planeside is better for passengers."
- What about boarding the airplane on the ramp, outside in the elements? "Not having a jet bridge is a lot faster."

And yet . . . even Cohen concedes: "It may not be identical in all respects."

But no one can disagree with the statistics he proudly touts. The FAA's 2011 forecast projects passenger enplanements on regional airlines to rise from 170 million passengers in 2011 to 296 million passengers by 2031; during this same period the nation's regional fleet is expected to increase from 1,771 to 2,764 airplanes. Today even the most veteran aviation professionals are shocked to learn that 53 out of every 100 commercial flight departures in America are operated by regional partners. That's a staggering statistic, and by all estimates the number will continue to grow.

In fact, analysts estimate that in the twenty years to come, the only substantial growth in the U.S. aviation market will occur within the regional sector. Yet even these stats don't quite tell the story, because the United States is a "mature" market and this alleged growth actually just reflects a shell game wherein regionals are mostly poaching passengers from mainline airlines. As for keeping pace with such expansion, Cohen said, "One thing our [regional airline] members have demonstrated is an ability to adapt almost instantaneously."

That's a hard point to argue, considering that Chautauqua simultaneously operates on behalf of five different mainline airlines. It's worth noting that sometimes regionals serve major carriers without branding themselves as the mainline partner; for example, Gulfstream International Airlines operates for Continental Connection, but the aircraft's livery does not reflect the colors and logos of Continental. However, in most cases the partner's airplanes are painted to reflect the mainline carrier, which often adds to the confusion.

I told Cohen about my experiences on the FAAC, and how I advocated for greater transparency so that consumers will always be aware of the airlines they are booking. He replied that he considers transparency a fait acompli: "It couldn't be more transparent. You'd have to be blind or deaf to not know it." Yet in the context of how much commuter aircraft have improved in recent years, he also stated: "It's very rare that someone even knows they're on a regional aircraft these days." Seems like a textbook case of having and eating your cake: there is full transparency, and it's so effective that passengers don't even realize when they're flying a regional.

Over time I raised the transparency issue with dozens of experts, and I was struck by the equally strong opinions on both sides of the argument. When I told MIT's Swelbar about the many readers of *Consumer Reports* who write about confusion over regional codesharing, he responded, "I'm totally surprised by that. I'm struck that people are not aware. The more experienced traveler knows it." It is a critical enough issue that in October 2010 the NTSB held a symposium on "Airline Codesharing Arrangements and Their Role in Aviation Safety." Even so, Tom Haueter, the NTSB's director of the Office of Aviation Safety, maintains there should be no consumer perplexity: "They're all [Federal Aviation Regulations] Part 121 carriers. I get a little frustrated. Usually it's not totally secret. Do you really think American Airlines is operating such a small aircraft? There's always a little word there on the ticket. It's pretty clear."

Like so much else about commercial aviation, it's apparent there's a disconnect between how industry experts view the airlines and how passengers view the airlines. I came to agree with analyst Bob Mann, who says the issue is under the radar because that's how airlines have marketed regional partnerships.[1] Millions of American passengers have no idea what carrier they are booking when they buy a ticket on a regional.

The DOT Inspector General's Office has been fighting for more detailed safety reporting from regional airlines in the wake of the Buffalo crash, and a source told me that some regional airlines don't have the resources to do this, so their mainline partners or the FAA must assist. He also thinks that in many cases regional aircraft livery—fuselage paint jobs—are deceptively reminiscent of the major carriers' livery.

Michael Levine, the law professor who was a senior executive at several U.S. airlines, sums it up: "I'm for identifying the operator. If Delta wants to brand it, then Delta needs to be responsible for how well it is operated. But Delta should be permitted to brand it and shouldn't have to operate it." Unfortunately, I don't think this is enough. If mainline carriers are selling tickets on regionals, they should be responsible for ensuring the safety of these operations.

Further complicating matters is that, yes, there are rules in place for revealing the operators of regional carriers when consumers book such flights. But such rules are often violated. Proof came in December 2010, when the Families of Flight 3407 issued a statement calling for full disclosure of every regional carrier operating flight segments on a given itinerary. This followed a report in the *Buffalo News* that numerous online ticket sites—Cheapair, Expedia, Kayak, Priceline, Travelocity, and TripAdvisor—as well as USAirways.com make it "difficult if not impossible for its customers to determine this information."[2]

That's why we at Consumers Union were happy to report in January 2011, one month after we submitted our final recommendations to the FAAC, that the DOT announced a new ruling that requires airlines and other online ticket sellers to disclose codesharing agreements on the same travel booking screens, next to all itineraries. DOT secretary LaHood stated: "When passengers buy an airline ticket, they have the right to know which airline will be operating their flight. For years we've required airlines to inform consumers about codesharing arrangements, and we'll be monitoring the industry closely to make sure they comply with the provisions of the new legislation."

But despite these best efforts, confusion over regional codesharing still reigns in some quarters. And it's not helped by the

DOT itself, or its FAA subsidiary. The record for an FAAC meeting in October 2010 in Denver reflects my concerns: "Mr. McGee stated that a similar issue exists with FAA safety statistics. He noted that family members of passengers killed in accidents involving regional operators flying on behalf of legacy carriers have taken offense at FAA and DOT statements that there have been no fatal major air carrier accidents in the United States since 2001."

Karen Eckert calls such announcements "wordsmithing" and says, "It's meant to detract and distract from the business that needs to be done. You see, even the next [regional] accident will take fifty lives and not three hundred and fifty lives, so they've got cover. It's insured."

Service and Performance Lag Behind the Big Airlines

For those who believe regional airlines offer worse service than mainlines, empirical evidence confirms that's not a misconception. An examination of DOT statistics for 2010 shows that in nearly all the major service categories, regional airlines performed worse than their mainline counterparts, and often performed much worse.

For example, in on-time performance, most regionals were in the second tier, and Comair ranked dead last with an on-time rate of just 73.1 percent; four of the top five carriers for canceling flights were regionals as well. As for mishandling baggage, regionals filled the seven lowest spots in the rankings of the eighteen largest airlines. American Eagle, which ranked eighteenth, had a mishandled baggage rate more than four times greater than top-ranked AirTran. And when it came to involuntarily bumping passengers, regionals once again occupied the bottom of the rankings, with Mesa and American Eagle far outpacing the remaining sixteen carriers in this category. The regionals fared better only in the consumer complaints category, with most ranking near the top, above their mainline partners. Ironically, it's not clear if this is due in part to complaints being filed against mainlines *because* of their regional partners.

Pat Friend of the Association of Flight Attendants, my fellow FAAC member, says regional flying is all about economics: "They call it rightsizing. Nothing stops [mainlines] from buying small planes—but they don't do that." On the very week of the terrorist attacks in September 2001, airline labor leaders were addressing the salary differential—what they termed "bridging the gap"—between legacy carriers and their regional partners. Friend adds, "That still is our goal, to bridge this gap. Our argument is that the job is the same regardless of the size of the airplane." In fact, for flight attendants working regional routes, the job can be even tougher, since only a single attendant is often assigned to such small aircraft, and taller crew members are forced to slouch.

Aviation law professor Paul Dempsey asserts that all this workplace feuding is ultimately felt by passengers and denigrates the product, as passengers are being funneled into smaller and smaller aircraft. Or as columnist Joe Brancatelli puts it, "If you have to fly a route with fifty-seat airplanes then you're unprofitable. You're diluting your brand."

Onboard Comfort and Amenities Don't Measure Up

Back in chapter 1, Jami Counter of TripAdvisor used SeatGuru.com standards to analyze airline seating comfort, but I also asked him how rankings of airline seating comfort extend to regional carriers. He immediately responded: "Everything I've just said does not apply." In fact, Counter acknowledged the regional airline product is "inferior" to its mainline counterpart: "I hate to say it, but the airlines have been able to game the system. They've glossed over it from a marketing perspective."

Strong words from an airline seating expert. But whether it's seats that are not as wide or seats that offer much less legroom or cabins that require passengers of average height to stoop, there's no denying that mainline airplane cabins and regional airplane cabins are apples and oranges—or apples and grapes. On one of Alaska Airlines' commuter partners, pitch is down to just 29 inches and the width is only 17 inches, woefully inadequate for the average American butt.

There's also the issue of smaller or even nonexistent overhead bins, so even carry-on baggage must be checked at the gate, thereby increasing the chances of mishandling and delaying the total flight experience. As noted, regionals often require outside boarding on portable stairs, even in wet and oppressive weather of all kinds. Smaller planes are also much noisier, and for those subject to airsickness, much more susceptible to turbulence as well. In fact, many regional aircraft are not equipped to fly above bad weather the way mainline jets often do. A SAAB 340 or deHavilland Dash-8, both operated by US Airways Express, provide bumpier rides than even the smallest of jets.

Yet another issue is child safety restraints. They are sometimes incompatible on commercial aircraft, and this is a particularly acute problem on regional planes. This was demonstrated in December 2010 when the mother of a one-year-old was booted off a United flight in Aspen operated by SkyWest when her safety seat didn't fit.

Of course, many regional planes don't have galleys. So when you consider that the average business-class and first-class passengers on mainline jets are offered unlimited drinks, hot meals, sleeper seats, and an array of onboard entertainment choices, it's mystifying to consider that passengers pay more for "premium class" on regional jets, where the price can run five times as much. "I don't think you can justify the cost domestically," says Counter, asserting that nearly everyone flying in first class on a regional is not paying for it. He explains that most are upgrading for free using frequent flyer mileage, or connecting to or from long-haul flights.

Yet even redeeming frequent flyer miles is impacted when passengers fly on regional aircraft. When airlines cut service on a given route, they often downgrade from larger to smaller aircraft, but regional jets often don't offer first class, and the front of the plane is a much less comfortable experience. And in a very ironic twist, low-cost carriers such as JetBlue and Virgin America that fly larger jets suddenly look much better than the majors, when those majors are outsourcing flying to regional jets. It's a competitive advantage, as evidenced by JetBlue boasting front and center on its website of spacious economy-class legroom.[3]

But cabin and seat size issues can transcend comfort. For many passengers, especially the elderly, oversized, or handicapped, regional aircraft simply aren't suitable. This has been borne out through the years when the DOT has fined regional carriers for not adhering to the Air Carrier Access Act of 1986, which "requires airlines to provide assistance to passengers with disabilities in boarding and deplaning aircraft, including the use of wheelchairs, ramps, mechanical lifts, or service personnel where needed." Little wonder that Atlantic Southeast Airlines was assessed a $200,000 civil penalty by the DOT in July 2011 for a number of violations. Considering the lack of jet bridges and the cramped confines on many regional flights, it's surprising there aren't more disability complaints filed with the DOT.

Congestion Issues: Smaller Planes Clog the Skies

To gain some perspective on just how pervasive regional partnerships have become, it helps to examine a major airline's total operation. For example, US Airways operates 3,310 daily flights, but only 1,291 of those departures are "mainline" while a whopping 2,019 are regionals. Also, the carrier has 338 aircraft in its fleet, but its nine regional partners fly another 303 planes. And for sheer size and scope, American Eagle (which calls itself "the largest regional airline system in the world") is larger than many mainline airlines, operating more than 1,700 daily flights to more than 150 cities throughout the United States, Canada, the Caribbean, and Mexico.

What some would call unnecessary congestion is a source of pride for the Regional Airline Association; Cohen notes that at the beginning of each day, in the hours between 5 a.m. and 9 a.m., there are thirty-five northbound nonstop flights between the three Washington, D.C.–area airports and four New York City–area airports—and all but five are operated by regionals.

For many airline executives, regionals are the new reality, and such outsourcing will only increase. Cohen explains: "From my first day in the industry, I was always told we were to throw more asses into seats. Don't you think the mainlines would do that if they thought they could?" When I asked how the nation can absorb all this airborne congestion, his answer was quick: "Fix the air traffic control system."

The answer to that, of course, is the fabled NextGen air traffic control system, the technological upgrade the DOT is seeking for the nation's outmoded current network. But a key sticking point in NextGen's implementation has been, Who will pay to equip the aircraft—the airlines or the FAA? And if the airlines are going to be responsible for making expensive high-tech investments in satellite technology, the deep-pocketed mainlines seem better able to ante up than the smaller regional companies. Yet Cohen predicts that in general regionals will be equipped just like their big brother partners, and he says that they will make these investments on their own.

The promise of saturated flight frequencies—not just a 2 p.m. departure, but a 1 p.m. and a 3 p.m. as well—clearly has left most mainline airlines quite overextended. The airlines are selling what customers want, not what they can deliver. And there may be more method than madness to such strategies; some industry experts believe the real motivation behind the major airlines' madness is to hog space at congested airports and prevent rival carriers from operating competing flights.[4]

That's why consumers are given only half the bargain: yes, there are more flights, but they are coupled with more delays (not to mention higher ticket prices). Fewer frequencies but improved on-time records would be a better choice.

All this congestion generated by sending more and more small airplanes into the air could not come at a worse time in the history of American commercial aviation. Busy airports such as LaGuardia and JFK in New York City, National in Washington, and O'Hare in Chicago are operating above capacity on a daily basis, and entire sections of the country are affected when delays start stacking up at the major hubs.

What's worse, smaller regional airplanes do more than just occupy space that could be allocated to larger jets; in fact, they slow down the entire system even more. Regulations require that different sized aircraft in the air traffic flow be afforded additional "miles in trail" separation for takeoffs, cruising, and landings, thereby generating even more delays. The in-trail requirement for back-to-back large and small aircraft is about six miles, nearly double what is needed for two large aircraft. Not to mention that the current system is a regression, since in the 1980s many airlines utilized airports that allowed for smaller turboprop aircraft to operate on shorter parallel runways, thus freeing up space both on the ground and in the air, a practice that has become much less common.

The only good news is that regional jets have gotten larger in recent years. According to the RAA, when deregulation was launched in 1978, the average seating capacity on a regional aircraft was 16 passengers and the average flight distance was 129 miles. By 2011 that number had risen to 56, and the flight length had more than tripled to 457 miles.

Furthermore, by 2030 the average regional aircraft is expected to hold 65 seats, meaning the old days of noisy turboprops are passing, and the mainline airlines will continue outsourcing more of their short-haul flying than ever. The FAA's twenty-year forecast for 2011 predicts an average annual growth rate of 4.1 percent for regional airlines every year through 2031, and that's in the most mature of all markets, the United States. So in actuality this "growth" is not growth at all; it's just a transfer of wealth from real airlines to their silent surrogate partners. For many passengers, there will be no choice.

Regional Airplanes Accelerate Climate Change

More airplanes in the sky mean more carbon emissions in the sky. And that is the problem. I happen to be very familiar with this phenomenon because I used to release dozens of aircraft in the busy Northeast corridor when I was an operations control duty manager for the Pan Am Shuttle. One of the primary functions of airline dispatching is to ensure that every flight operates as economically as possible by burning as little fuel as possible. So while that was always a financial consideration, I'm neither proud nor ashamed to state that in the early 1990s none of us ever considered aircraft fuel burn in the context of the environment, only in the context of cost savings. It literally was not on our radar.

Juan Alonso, the aeronautical environmental expert at Stanford University who served beside me on the FAAC, maintains that so much additional regional flying made sense back when fuel was cheap enough. Now it's not only harmful to the planet but too expensive as well. He suggests high-speed rail should be the dominant mode for trips up to five hundred miles.

But there is a problem particular to regional fleets: empirical evidence that smaller aircraft are less efficient and spend more time on climb.[5] Many of us may view such issues in terms of automobiles: that is, small cars are better than large cars because they burn less fuel. But when it comes to mass transportation, the environmental sweet spot is often contained within the largest vehicles, even if they are aircraft with four jet engines. It's all about moving the most people the most efficiently. Experts say the current practice of operating three to six regional aircraft on routes that could be serviced by just one large aircraft is a formula that is killing us softly.

At my meeting with Secretary LaHood, I asked about all those planes crowding the skies. He responded, "People are making those choices every day here in the Northeast Corridor. That's why the trains are full, that's why Amtrak is making money, that's why ridership on Amtrak is way up. But there will always be people who will want to ride a plane because they just will and they believe they will get there quicker. The point is, we can do both. But in the end I think people will always decide on air travel."

With Regional Airlines, Safety Questions Abound

For many passengers, this may all boil down to—So what's in a name, anyway? Whether it's Delta or Chautauqua, who cares what company is listed on the FAA's operating certificate so long as we arrive on time? Fair enough. But what if you don't arrive at all, as was the case with five codeshared regional flights in recent years? Because we've learned there's a world of difference in how Continental Airlines operates and how Colgan Air operates.

During the 1990s there was a string of fatal regional airline accidents, at a time when "commuter carriers" were governed by Part 135, a separate section of the Federal Aviation Regulations. A report from the U.S. General Accounting Office in 1992 noted the increasing accident rate—twenty-two in 1991 alone—and cited "inadequate inspections" by the FAA. Pressure from Congress, labor unions, and passenger advocates helped convince the DOT that all commercial aircraft with more than ten seats operating scheduled passenger service should adhere to Part 121, and the change was made in 1995. As an FAA-licensed dispatcher and former flight operations manager, I remember the relief I felt at that time, knowing that until then regional airlines were not even required to hire dispatchers, a federal mandate since 1938 for larger carriers.

So theoretically, for the last seventeen years there has been one safety standard for both mainline and regional airlines in the United States. But many experts maintain that in the real world regional airlines de facto adhere to lesser standards for pilot qualifications, crew training, and operational and maintenance requirements. The Buffalo crash made this painfully clear. The fact is that some of these partners are the weak links for the mainlines, in the most literal sense of that term.

The Professional Aviation Safety Specialists, the union that represents FAA inspectors, has been vocal about regional airline safety for years. When I pointed out the multiple fatal accidents involving U.S. regional carriers over the past decade, PASS president Tom Brantley said, "The FAA attitude is those don't count."

He said that real differences still exist: For one thing, DOT safety statistics are kept separately, so a true big picture of U.S. airline safety is not presented. What's more, training for regional carriers allows pilots to continue operating after failing a specific function during simulator training, while mainline pilots must master such tasks before returning to the line. And as the Colgan accident made clear, crew rest and commuting issues seem more prevalent among regional airlines.

As for security, what many passengers don't realize is that some regional carriers have been given waivers from cockpit door barriers—there not only is no reinforcement on the cockpit door, there simply isn't a cockpit door *at all* on certain smaller airplanes. This was made painfully obvious in 2006 when a passenger made his way onto the flight deck as a Beech 1900-D operated by Gulfstream International Airlines—which codeshares on behalf of Continental—was taxiing on an active runway.

There are inherent safety risks in flying regional carriers. One is the route maps of the regionals, because they often fly into smaller airports not serviced by legacy carriers, and this can mean fewer or less advanced aeronautical navigational aids. It can also mean second-tier crash and rescue protection. What's more, the physics of smaller aircraft make survival statistically less likely in a regional airplane than in a larger aircraft.[6]

Time and again, discussions about regional safety return to Flight 3407 in Buffalo. As MIT's Swelbar notes, "Colgan taught us that not all regionals are created equal." He and others cite SkyWest, Pinnacle, and Republic as the strongest regional carriers, but experts claim that Colgan and Mesa are not comparable companies.

That's why Bill Voss of the Flight Safety Foundation asserts that ISO certification bestowed by the International Organization for Standardization—a Geneva-based organization that represents 159 countries around the world—is a tougher standard than the FAA's. He says, "It's that mix-up that compliance correlates to safety—and it doesn't. You have A students and A+ students if you grade them under ISO standards. There are several U.S. regionals that would not meet the ISO standards. If I were head of the Regional Airline Association, I'd have a smaller club with the cream of the crop." One former major airline executive agrees with this sentiment, and told me his company would not work with Mesa.

And the crew fatigue problems highlighted in the Buffalo crash continue to reverberate throughout the industry. "One thing that's never discussed is pilot commuting," says Bob Crandall, the fiery former CEO of American Airlines. "They live on the other side of the country or they live outside the United States, and they commute to fly. And then want to talk about fatigue in the cockpit. Well, you also have to talk about all the flying you did before you signed in for duty. . . . But the problem is the FAA is not doing as good a job as it should because it has not done anything about circumstances like the Colgan situation."

But pilot pay and pilot commuting are linked, and few organizations have done more to highlight such inequities than the National Air Disaster Alliance/Foundation, which assists crash victims' families, so I asked the executive director if things have improved since the Buffalo accident. "Nothing has changed," Gail Dunham said flatly. "It's an enormous disappointment. It's still about pathetic pilot pay. The pay scale still starts at seventeen thousand dollars a year. Did you ever think you would get on a plane where the pilot is making seventeen thousand dollars a year? It's appalling."

When I asked Cohen of the Regional Airline Association about the wide disparity between crew salaries for mainline and regional airlines, he said, "It doesn't affect the product one bit." Cohen explained: "Since the Wright brothers, salary has been linked to two things: the size of the aircraft and seniority. And contracts are freely collectively bargained by both parties." He notes that in February 2011, 90.5 percent of the pilots at Pinnacle, Mesaba, and Colgan ratified a new five-year contract.

In the wake of the Buffalo accident, there's no question the salary issue caught the public's attention. But Voss maintains the problem runs deeper than pilot pay rates: "There's no incentive in the system to develop those pilots because they're going to come in and go out. If something is fundamentally wrong, it's that there are two different types of regionals. If they are going to be minor-league teams, then treat them like minor-league teams. But it's a race to the bottom, so you see two radically different models."

Salt in the wound for Flight 3407 family members was the testimony of Jeffery Smisek, then the CEO of Continental and now the CEO of United. He responded to congressional inquiries into the Buffalo crash by suggesting it was not Continental's duty to oversee the safety of Colgan's operation, and stated: "That is the responsibility of the Federal Aviation Administration." The Eckert sisters were present that day, and when I brought this up with Karen, she responded, "Oh, that was horrible. It was like a punch."

Other airline executives whisper off the record that Smisek screwed up that day—big-time—but rival airlines have taken the Smisek approach as well. When asked about a regional partner by the *New York Times*, a US Airways spokesman was quoted as saying, "We've got our own airline to run." And at a Senate hearing in 2009, a senior vice president at Delta stated: "We do not, and cannot, directly manage our regional partners' safety and quality issues."

So do some mainline airlines take more responsibility for their regional partners than others? It's a topic I've raised with both current and former airline executives, and responses vary. Take Glenn Tilton. The CEO of United and I crossed swords on many issues during FAAC meetings, but in the end we fundamentally agree on the need for mainline airlines to assume responsibility for their regional partners. He explains that United (now merged with Continental—ironically enough, considering Continental's recent history) implements a "two-pronged policy" of constantly exchanging best practices on safety and reliability and launching thoroughly objective audits of regional operations, with the price of failure being termination.

Yet the executive in charge of the Regional Airline Association asserts that the safety burden remains with the regionals. When I asked Cohen about Smisek's congressional testimony, he said, "The legal and regulatory responsibility is on the operator." But he added, "The other half of the equation is that there's one level of safety and has to be. . . . The most fundamental element is trust. If you're a mainline airline, you're entrusting the most sacred things: the safety of your passengers and the value of your brand. The frequent flyer program, the cocktail napkins, everything—it's all branded."

Tilton, however, dismisses the proposal that legacy airlines assume legal or operational control of their regional partners: "I don't think that's possible with publicly traded companies." One thing is certain: the current tangled web of multiple mainline/regional partnerships undoubtedly would make such rules difficult to implement.

Some context is required here. The dirty little secret among airline executives is that they believe labor—and particularly labor unions—"forced" the major carriers into outsourcing as a way of reducing costs. And this applies to call centers, to maintenance repair shops, and to regional partnerships. When I met with Crandall and asked about the explosion of regional airlines, he told me: "That, of course, is all a reaction to labor."

Airline veteran Michael Levine believes most mainline airlines don't want to own their regional partner airlines, in order to avoid the stress and cost of labor contracts. I think he's right, but the end result is that they've created a second-tier regional system with lower pay scales, and therefore created lower-quality service. However, Levine blames the unions for this situation, while I believe that airline executives have allowed service and standards to deteriorate as wages have fallen.[7]

There's ample evidence of poor management of regionals. A core group of Flight 3407 family members has made more than forty trips to Washington, which Karen Eckert explains thus: "We have data. We don't just go and cry. We're hoping to raise the consciousness so there is a difference." They've had an impressive impact: they showed me personalized letters from members of Congress, and in May 2010 President Obama met with ten family members in Buffalo to hear their stories. But their greatest accomplishment came in August 2010, when Obama signed into law the Airline Safety and Federal Aviation Administration Extension Act of 2010. Among the key provisions are strengthening standards for pilot training, enhancing FAA oversight of regional carriers, and increasing transparency for codeshared flights. Unfortunately, critics claim it will not do enough.

Mainline or regional, it's all about FAA oversight. After her sister's death on Flight 3407, Susan Eckert Bourque learned how such marketing agreements work: "When it suits them, the regionals are independent. When it doesn't suit them, they're not. They kept saying, 'It's the FAA's job.'" She cites the need for stricter oversight of regionals and adds, "It has to be done through legislation and regulation. We can't trust the airlines to police themselves."

Last year I asked Randy Babbitt, then the administrator of the FAA, if pilot proficiency at the regionals currently is up to par, and his response was intriguing: "We make a set of rules on training standards. And if you want to do it to a higher standard—that's your privilege. Some are going well beyond." He also noted that many mainline carriers provide state-of-the-art flight simulators in training. In other words, bigger airlines with bigger pockets exceed the regulations, while smaller regional airlines often can't compete.

I went further, and bluntly asked Babbitt if all U.S. regional airlines are compliant with the FARs. He said, "Absolutely. We oversee fifty thousand flights a day and fifty-five hundred commercial aircraft. We have seven thousand inspectors but we can't monitor all the flights. That's why we use spot checks, data, trends, and designees." So perhaps the root problem with regional carrier oversight—and maintenance oversight as well—is not about adhering to the FARs, but about strengthening those regulations? That's the inescapable conclusion I drew.

Without prompting, Babbitt's boss referenced the Buffalo catastrophe as well. When I met with Secretary LaHood, I asked about the challenges since taking office in 2009, and he responded, "I think the saddest day for all of us was when the Colgan Air plane crashed in Buffalo on that cold winter night and fifty people perished. But we stepped up immediately. Randy traveled the country and he had ten or twelve safety summits. We immediately put new enforcement rules in place for rest for pilots and training for pilots. Because we know there were flaws in the operation of that particular flight. The pilots were not well trained. They weren't well rested. And we learned some very, very tough lessons from that tragedy."

Those words resonate when you visit the tiny town of Clarence Center, a hamlet within landing gear distance of Buffalo Niagara International Airport, and you stand on a street where screaming metal once fell from the cold sky. In an age when technology has brought an immeasurable margin of safety to the art and science of flying, these were fifty deaths attributable to the oldest of demons: human error. But even that assessment does not tell the full tale, for ultimately the two Colgan Air crew members were victims as well. Human greed, though it may be harder for the NTSB to categorize, is a much deeper and indelible systemic cause. The regional airline industry's hiring practices, the salary scales, the training programs—all this can be traced back to major airlines that seek to place more and more human lives in the hands of the lowest bidders.

Back in 2003, Beverly Eckert wrote a column for *USA Today* that was at once stirring and eerily prescient: "My Silence Cannot Be Bought." She detailed her decision to refuse a "$1.8 million average award" payoff for her husband's death on 9/11 and instead chose to file suit to learn the real causes of that day's horror: "Lawmakers capped the liability of the airlines at the behest of lobbyists who descended on Washington while the Sept. 11th fires still smoldered." And she closed by stating: "So I say to Congress, big business, and everyone who conspired to divert attention from government and private-sector failures: My husband's life was priceless, and I will not let his death be meaningless. My silence cannot be bought."

Unfortunately, in the wake of her own fiery death, and plagued by a new set of unanswered questions, Beverly Eckert's challenge to speak out against government and private-sector failures still looms.

7

Outrageous Outsourcing: The Single Greatest Threat to Airline Safety

The machine does not isolate us from the great problems of nature but plunges us more deeply into them.

—*Antoine de Saint-Exupéry, pilot and author*

TWENTY-ONE FATALITIES—BUT NOTHING MAJOR. THAT'S how the airline industry views the one accident that has stood out from all others because it foretold the dangers to come. This crash remains the wake-up call the airlines *still* refuse to acknowledge.

In January 2003, Flight 5481 took off from Charlotte, North Carolina, en route to Greenville, South Carolina. The Beechcraft 1900D, operated by Air Midwest on behalf of US Airways Express, had barely gotten airborne when it stalled and crashed into a US Airways hangar and burst into flames, just thirty-seven seconds after leaving the ground. All twenty-one persons on board were killed and one person in the hangar was injured.

The NTSB determined that the probable cause was the airplane's loss of pitch control during takeoff, which had resulted from the incorrect rigging of the critical elevator control system by an outsourced maintenance contactor, Raytheon Aerospace in Huntington, West Virginia. In turn, Raytheon had further outsourced some work to unlicensed maintenance contractors—which, incredibly enough, is *not* a violation of FAA regulations. Whose fault was the crash? As it often does, the NTSB cited multiple parties. These included Raytheon, Air Midwest (for its "lack of oversight of the work being performed at the Huntington maintenance station"), and the FAA itself (for lack of oversight of Air Midwest's maintenance program).

So what about US Airways? Well, even though the doomed airplane was painted in that airline's colors, federal authorities are blind to such nuances. This was strictly an Air Midwest tragedy, not a US Airways tragedy.

As airline accident investigations go, this one was fairly routine, inasmuch as such heartache can ever become routine. The NTSB wrapped up its findings in little more than a year. But then a funny thing happened. Word got out that some of the family members of the victims—specifically the parents of Christiana Shepherd, an eighteen-year-old college student—had not accepted an offer of settlement from Air Midwest's insurance reps. Instead the Shepherds wanted a public apology. The attorneys conference-called and conference-called, and in a nation where a dollar sign can be affixed to anything or anyone, the airline's lawyers began to get a little frustrated. Just what exactly did these parents want? An apology. No, really. How much? They want an apology. No, come on. Give us a figure. That *is* what they want—an apology. And so it went. Nearly four years after the accident, I spoke to Pastor Douglas Shepherd and his wife, Tereasa Shepherd, for *Consumer Reports.* "The bottom line for these [airline] companies cannot be money," Christiana's father told me. "That has to change. Outsourcing with no oversight leads to loss."

It took more than two years, but in the end, the Shepherds caused an extraordinary event to take place, something unique in U.S. commercial aviation. The head of an airline attended a memorial service, stood before the next of kin, and read a formal and public apology. On May 6, 2005, Greg Stephens, the president of Air Midwest, joined victims' families at the crash site memorial in Charlotte and uttered these words: "We are truly sorry, and regret and apologize to everyone affected by this tragic event."

In the interim, the outsourced maintenance company had—you guessed it—changed its name, from Raytheon Aerospace to Vertex Aerospace. But even that new entity was invoked by Stephens in his remarks: "Air Midwest and its maintenance provider, Vertex, acknowledge deficiencies, which together with the wording of the aircraft maintenance manuals, contributed to this accident."

Obviously much credit goes to the Shepherds for making this happen. And credit where it's due to Air Midwest for breaking the nearly century-old tradition of airlines invoking a virtual curtain of silence after a tragedy. But a few qualifications are required: Soon after the apology, Vertex Aerospace changed its name *again*, this time to L-3 Communications Aerotech. Then Air Midwest was shut down by parent company Mesa Air Group in 2008. And despite the eloquence of Stephens's apology, at no time did he mention the airline whose name was emblazoned on the side of the defective plane: US Airways.

Cutting Corners

Despite tremendous advances in technology, aircraft maintenance remains a critical threat to airline safety. Many passengers may be surprised to learn that airline accidents caused by maintenance factors have increased significantly in recent years. In fact, PlaneCrashInfo.org has categorized causes of fatal commercial accidents worldwide from the 1950s through the 2000s, and the percentage of accidents due to mechanical failure between 2000 and 2009 was 28 percent, higher than in any of the previous *five decades.* That total was second only to pilot error.

There's a tasteless joke that's been kicking around the industry for years: "If they ever find a way to kill the passengers without killing the pilots, then the industry is in trouble." What's implied, of course, is that crew members are putting their own lives on the line when they take off in an airplane with a questionable airworthiness record, but even so there are limits. The military addresses this issue by often requiring mechanics to fly in the aircraft they service. But I can testify under oath that every day airline pilots take up aircraft they know are lacking in critical maintenance; I saw it firsthand at the small carriers where I worked.

"That ninety-nine-dollar ticket came with a price," a mechanic for a major U.S. airline told me. "And the savings came from maintenance." This isn't hyperbole. The industry's collective decision to outsource maintenance has become a threat to everyone who flies commercially these days. Among domestic carriers, there is only one exception: American Airlines continues to perform nearly all maintenance in-house, but the company's recent bankruptcy filing may change that policy.

Airline executives, FAA administrators, and Wall Street cheerleaders all maintain the same drumbeat: wrenches are wrenches, so it doesn't matter in which country they are turned, nor does it matter whether the hands turning those wrenches are working for a living wage or a subminimum wage. Why should a passenger care if an airline pays an outside company to hire a mechanic who will work for less than that airline's own mechanics? Financial analysts are quick to remind us all how greedy those union leaders are (though they dare not criticize the exponentially more excessive greed of airline executives). If the work is done properly, and the airline saves a few pennies per hour, then it's a win-win.

I asked Tom Brantley of Professional Aviation Safety Specialists what I have been asked by so many skeptics—editors, fact-checkers, interviewers, attorneys, even passengers: If things are so bad, why aren't planes falling out of the sky? He responded, "Is that really the criteria we want? Once a plane falls out of the sky, is that when we change? It's like a relative with a drug problem. That person is going to crash and burn but may look okay." Brantley added, "I think there's a margin of safety. And that margin is razor thin now. There are no more buffers. There is no buffer zone now. The FAA doesn't know where the margin is." FAA whistle-blower Gabe Bruno says, "That's exactly where the problem is—the FAA is waiting for a fatal accident. That's the wrong benchmark. If there's no accident, it won't get picked up at all."

Former FAA administrator Randy Babbitt acknowledged the concerns I raised in the FAAC and said, "It's a challenge, there's no question. There is greater and greater reliance on power by the hour [outsourced maintenance agreements]. . . . We've had recent incidents that were the fault of outsourced maintenance." But he asserted some foreign governments are assisting the FAA with surveillance: "We have confidence with certain regimes, for example, the European Union. Whatever oversight is provided, we trust their inspectors are going there."

Today most passengers have no idea how dangerous the outsourcing paradigm has become. Consider:

- Although regulations indicate all maintenance shops are treated equally, in reality there are two models, and independent facilities can differ on hiring, training, security, and drug and alcohol screening.
- In addition, there are two sets of rules for certificated and

noncertificated repair shops, and more work is being *sub*-subcontracted to noncertificated shops.

- Many outside contractors are not FAA-licensed mechanics; in some cases, the technicians cannot read aircraft manuals in English.
- The most critical problem of all—documented repeatedly by several government agencies—is the FAA's inability to provide consistent oversight of outsourced maintenance, both in the United States and abroad.

The airlines are engaged in a "mad race to the bottom on costs," according to Kevin Mitchell, chairman of the Business Travel Coalition. "You can't win this argument on economic grounds," he says. "Yes, there are savings in outsourcing maintenance, but it shouldn't screw up your bigger picture analysis." He also notes how misleading statistics can be: "In an era of great change, statistics are not as predictive as over a forty-year normalized period. I would debunk this argument by looking at Deepwater Horizon. The day before the explosion, there had been zero mishaps. But there had been signs it could happen."

Some historical perspective is critical. U.S. commercial aviation was the world's gold standard for nearly ninety years, but times have rapidly changed, and too many laurels are being rested upon. What has occurred over the last ten years is a complete break with industry protocol. On the FAA's website are the locations of many of its outposts, termed Flight Standards District Offices (FSDOs), widely known as "Fizz-does." For decades, the FAA put its resources where the airlines put theirs. So the FAA's principal maintenance inspector for Delta would be located in Atlanta, just down the road from Delta's primary maintenance facility. So too for Northwest in Minneapolis, American in Tulsa, etc. For generations it meant that at any time of day or night an FAA inspector could pop in unannounced and watch as those wrenches were turned. Veteran FAA folks call it kicking the tires.

Today those FSDOs are still in the same cities, but the maintenance work has moved elsewhere, sometimes to the other side of the world, leaving scores of FAA inspectors behind with little to do but inspect empty hangars because politics and budgets have prevented them from doing their jobs.

The new outside repair stations have names such as TIMCO, HEICO, SASCO, Sonico, Ameco Beijing, Aeroman, Hong Kong Aircraft Engineering, MTU, and ST/Aerospace. They're located throughout the United States and in Europe, as well as Mexico, El Salvador, China, and Singapore. Unfortunately, in many cases airplanes come back from such facilities in need of more work.

David Bourne is the director of the Airline Division of the Teamsters, but he is also a veteran pilot qualified in multiple aircraft types. Therefore his take is unique: "This is not a paper argument. I have personally encountered problems with outsourced maintenance. We could tell where the repairs had been done when we flew the planes after servicing."

Take the tragic crash of Air France's supersonic Concorde in Paris back in July 2000, a disaster that claimed 113 lives. The image of that damaged aircraft spewing plumes of fire, its fuel tank aflame upon takeoff, was transmitted around the world. That photo hastened the end for the entire Concorde fleet; both Air France and British Airways grounded them for good by 2003, thus ending commercial supersonic transport. What was not as widely reported, however, were the subsequent findings of French authorities that the fire (and resulting crash) was caused by a tire that ruptured just as the plane accelerated for takeoff, after that tire struck a piece of titanium debris on the runway. And the debris? It turned out to be a wear strip from an engine cowling that had just fallen off the last plane to depart from the same runway, a Continental Airlines DC-10. And *that* aircraft, in turn, had just been serviced by an outside contractor, Israel Aircraft Industries in Tel Aviv. In December 2010, a French court found Continental and one of its employees criminally responsible.

How bad has the maintenance outsourcing problem gotten for U.S. airlines? An independent analysis by *USA Today* found that between 2004 and 2009, millions of passengers flew on 65,000 flights operated by domestic carriers that should never have left the ground. That's right, a total of *65,000 flights*.

Here's a small sampling of troubling indicators:

- In 2008, a Boeing 737 operated by United declared an emergency air turnback to Denver due to a low fuel pressure indicator on one of its two engines. The reason? Rather than using the proper protective caps, mechanics had used two shop towels to cover the oil sump covers on that engine.
- In 2009, the FAA imposed a $5.4 million fine against US Airways for operating three airplanes after the FAA had issued Airworthiness Directives demanding the planes be grounded until inspected. Two of the planes were flown despite a warning about a possible crack in the landing gear and the other airplane was flown despite the threat of a cargo door opening in-flight.
- In 2010, the FAA levied a $348,000 fine against Chautauqua Airlines for operating eleven regional jets for more than 27,700 flights between 2007 and 2009 without performing required inspections. Further complicating the issue is that Chautauqua itself is an outsourced airline for five separate regional brands—AmericanConnection, Continental Express, Delta Connection, Midwest Connection, and US Airways Express—so it's unclear which passengers were in danger.
- In 2010, NPR broadcast a report about the Aeroman repair station in El Salvador, which was responsible for what was

termed "a mistake that could have potentially been catastrophic." In January 2009, a US Airways aircraft en route from Omaha to Phoenix was forced to land in Denver after the pressure seal around the main cabin door started to fail in-flight. Later it was discovered that Aeroman mechanics had installed a key component on the door *backward*.

An Outsourcing Outbreak

The latest numbers are hard to uncover, but there's no question that outsourcing keeps increasing. According to the Aeronautical Repair Station Association, the maintenance repair and overhaul (MRO) market annually exceeds $50 billion worldwide. Within the United States, ARSA claims the industry has grown to 200,000 employees at 4,200 firms, but small- or medium-sized enterprises constitute 85 percent of this total.

In September 2008 the DOT's inspector general issued a detailed report on aircraft maintenance outsourcing and noted that nine of the ten largest U.S. carriers were outsourcing 71 percent of their "heavy" maintenance and repairs, up dramatically from 34 percent in 2003. Also, 907 repair stations were performing heavy maintenance for the largest domestic carriers. Of these outside facilities, 661 were located in the United States, while 81 were in Central America, 79 in Asia, 69 in Canada, and 17 in Mexico. (The United States and the European Union have a reciprocal oversight agreement—a "transatlantic NAFTA"—so European work is not included.) Overall, the FAA had certificated a total of 4,159 domestic and 709 foreign repair stations by July 2008, an avalanche of new outposts.

But even these numbers don't tell the whole story. Incredibly, regulations allow U.S. airlines to *voluntarily* provide such data.

Also, these voluntary requests cover only their "top ten substantial maintenance providers," even if an airline has hundreds. The IG found that reporting system to be "inadequate" and bluntly stated that the FAA "still does not have comprehensive data on how much and where outsourced maintenance is performed." There's no question the outsourcing percentages are climbing even higher today.

The IG redacted airline names, but among its other findings was that over a three-year period, FAA maintenance inspectors assigned to one airline visited only four of its fifteen most substantial maintenance providers. In another case, a major foreign engine repair facility was not visited by the FAA until *five years* after the FAA approved it for repairs; such an inspection should have occurred immediately. And one maintenance vendor continued performing work despite failing to respond to a total of *sixty* red-flag open corrective action reports that had exceeded the required response time windows. Yet despite agreeing to a specific recommendation in 2003, by 2007 the FAA still was not properly identifying 33 percent of outside repair stations in its database.

As for oversight, it's mostly the scout's honor system. The FAA relies heavily on air carriers' oversight, and audit programs are not always effective. The regulations are quite loose about when problems need to be addressed, and violations can be resolved simply based on written pledges from the outside repair stations! In fact, the FAA does not require its maintenance inspectors to conduct on-site inspections of outside repair facilities prior to or even *after* approval of those stations—even when problems are found.

What's more, there's plenty of blame to go around. The IG report stated: "At one heavy airframe repair station, *all three types of oversight failed*—FAA, air carrier, and repair station." In other words, the system is not working from top to bottom. In this case, two internal airline audits and two external FAA inspections failed to detect "significant weaknesses" that eventually shut down the outside facility for more than a month.

Loopholes, mushy language, omissions, communication gaffes. The regulations are Swiss cheese, the airlines exploit them, and the FAA cannot or will not enforce what it clearly could enforce. In addition, there's a tremendous amount of undocumented work being done on the nation's airline fleets, which undermines the big-picture maintenance trend analysis continually conducted by the FAA, airlines, and manufacturers; in other words, if problems with a specific airplane aren't reported at Airline A, how can Airline B follow up?

Has any good come from all this scrutiny? Well, in January 2008 the FAA issued a directive requiring every U.S. airline to simply *list* all the certificated and noncertificated repair stations it uses. A positive first step at getting a handle on this mess, no? Bad news. The FAA promptly rescinded this policy, because the airline industry "expressed significant concerns" with the undue burden of reporting this.[1] Consider that. The FAA merely *asks* the airlines who is performing all this work and even just *listing* their vendors is too hard for Fortune 500 companies in the digital age. Couldn't someone just ask accounts payable where the checks are cut?

In fact, several government entities have been warning about the dangers of outsourcing for years, in report after report. Consider that in June 2005, the IG concluded: "FAA inspectors were not able to effectively use the oversight systems to monitor the rapidly occurring changes." Translation: electronic surveillance is being employed just when the airlines are offshoring tons of maintenance work. And in December 2005, the IG released a scathing report that concluded: "Neither FAA nor the air carriers [we reviewed] provided adequate oversight of the work that non-certificated facilities performed." It found that as many as 1,400 noncertificated repair facilities—including 104 foreign facilities—had *never* been inspected by the FAA. Translation: none needed. And in September 2006, the House Transportation Committee held hearings and noted that the FAA had addressed only one of nine IG recommendations made in 2003: "According to the IG, the FAA has not made much progress on those recommendations." Translation: what else is new?

The inspectors I've spoken to in recent years are based throughout the country. All of them are livid and most request anonymity, for fear of reprisals from FAA management. One said that "All inspections outside the U.S. must be coordinated with the State Department. Nothing is a surprise to them. A red carpet is laid out for you." Another reported that "Our inspectors just can't get [to Latin America]. Down there they've got twenty unlicensed mechanics with one licensed guy from the airline looking over them." Yet another told me, "You may have twelve to fifteen laborers working under the license of one mechanic. . . . It used to be if there were twelve mechanics working on an aircraft they were all licensed."

Dangerous Data Uncovered

Among the many programs the FAA oversees is the Service Difficulty Reporting site, whereby industry professionals are required to electronically log in all manner of discrepancies; eventually the SDRs are available to the public. But after a visit to Massachusetts, I was stunned to learn that the FAA's entire airline database is capturing only a small fraction of the serious incidents and events occurring in the airline industry every week. And a systemic problem with smoke in domestic airline cabins may be avoiding detection.

John King is a veteran mechanic who was licensed in 1963

and eventually fired by Eastern Air Lines as a whistle-blower in 1987 when he reported on that carrier's maintenance improprieties. I visited him and he supplied me with reams of data—so much so that I later told a former inspector that King is doing the work of ten people at the FAA, and the inspector corrected me: "One hundred people."

King's methodology is decidedly low-tech, but without remuneration he performs a critical and patriotic service by continually Googling key aviation terms: *diversions, emergency landings, flight returns, unscheduled landings.* He then gathers as much information as possible through news reports of all U.S. airline in-flight anomalies. Finally he periodically compares his private database with SDRs collected by the FAA. The results are stunning, and riddled with omissions:

- 2008: 112 of 148 incidents (76 percent) not filed; 132 of 148 incidents (89 percent) not in compliance
- 2009: 96 of 156 incidents (62 percent) not filed; 128 of 156 incidents (82 percent) not in compliance
- 2010: 118 of 183 incidents (64 percent) not filed; 153 of 183 incidents (84 percent) not in compliance

The compliance numbers are so low because additional incident reports either omitted a required cause or omitted required documentation (part numbers, etc.). But two other points are critical. One, King's database reflects only the forced landings and air turnbacks captured by news reports. Two, such diversions are but one symptom of a wide range of potential mechanical problems. In other words, as disturbing as these findings are, *they truly represent just a fraction of existing problems.*

Even more chilling, a significant number of these incidents were due to smoke in the cabin: 32 percent in 2008, 24 percent in 2009, and 26 percent in 2010. And overwhelmingly these events were not entered into the SDR database, with only 30 percent correctly filed in 2008 and 38 percent correctly filed in both 2009 and 2010. Consider that no major media outlet in the United States has reported on the fact that our nation's commercial aircraft fleet has had a widespread and systemic problem with smoke in airplane cabins—and the FAA has not been effectively tracking this problem.

"It's dysfunctional," says King of the SDRs. "It's financed by the government and it's the only mandated program. Without an accurate SDR database, there's no way to gauge the voluntary reporting programs." And he adds, "This doesn't cover a whole spectrum of other problems."

Last year, before he resigned following a DUI arrest, I asked Administrator Babbitt about the SDR system, and he responded: "We just want the data. We want a very concrete example of where we can identify that data." He cited the FAA's Aviation Safety Information Analysis and Sharing (ASIAS) system, and suggested, "Delta may not want United to know about its engine life."

But whistle-blower Bruno is not surprised by King's findings: "The margin of safety keeps getting worse. It's not a safety agency. . . . Today they are a facilitating agency, to facilitate the airlines' profits."

The rapid expansion of maintenance outsourcing introduced new risks into the FAA's safety oversight. These missing data underscore both a troubling increase in maintenance failures *and* deficiencies in FAA surveillance.

Double Standards?

At the FAAC, John Conley raised what came to be known as the "single standard." The administrative vice president for the

Transport Workers Union, Conley repeatedly asked if the FAA oversees all maintenance equally, whether in-house or outsourced, in the United States or overseas? FAA officials said yes. But the differences can be striking.

First, there's the issue of certifying mechanics with airframe and powerplant licenses. At one meeting, I noted that the historical model was for ten FAA-licensed mechanics to be supervised by one licensed supervisor, but now we've seen that at many outsourced facilities, ten *unlicensed* mechanics can be supervised by that same licensed mechanic. To the FAA, it's all about the certification—that is, the work is being certified by a licensed representative of the facility and/or airline. But common sense indicates these are two very different models. As Brantley says, "It's one thing if the people doing the work are licensed—but they're not. That is huge. Let's be honest—the only way to get your labor costs down is to use less qualified people." But Sarah MacLeod, executive director of the Aeronautical Repair Station Association (ARSA), dismisses the value of a license: "I know a lot of educated idiots. The certificate doesn't guarantee anything. . . . Will it not lead to an accident? We can't judge that."

Then there's drug and alcohol screening; the regulations require periodic testing of mechanics inside the United States, but this rule is waived overseas. At the FAAC, we were told that this is because America can't impose its values on a sovereign nation; true enough, but if such testing is illegal in another country, why is that country repairing U.S. aircraft? Bourne is dismissive of these arguments: "They say we can't make other countries do what we ask—well, we damn well do it with cars. If you want to sell cars in the U.S., you have to adhere to our rules." Meanwhile, ARSA has been waging a legal battle on this front since 2006, to prevent such regulations from being enforced at more than twelve thousand facilities nationwide. MacLeod says, "ARSA is not against drug testing, it's the level of testing."

And there are security concerns as well. "We were under armed guard for three days," says an airline mechanic who visited the Aeroman facility in San Salvador (servicing JetBlue, Southwest, and US Airways). "Guys with machine guns were guarding the parts bin. That's a security problem." A former air marshal calls the lack of security at maintenance facilities "a huge misdirection of resources." Bourne points out that while a sixty-year-old unionized ramp worker at Continental is subject to a background check that extends back to his sixteenth birthday, "bin Laden himself could have gotten a job fixing U.S. planes in El Salvador since the background check goes back one day."

Old-time mechanics recite an industry bromide: If there's a mechanical problem, you can't pull an airplane over to the side of the road. As for outsourced work, one mechanic for a major carrier says problems "run the gamut," citing improperly rigged cables, miswired and transposed cockpit instrumentation, and drill shavings trapped in wiring bundles. All could lead to catastrophic failures and fatal accidents.

Five years ago the Teamsters Aviation Mechanics Coalition was formed to "go on the offensive" over maintenance outsourcing. Chris Moore, a veteran mechanic for Continental, chairs the TAMC, which solicits input through Outsourcing Defect Reports, chronicling problems mechanics find on aircraft serviced by outside MROs. He says, "We've seen the airplanes coming back and we've been griping about the bad shape they're in."

Recently United mechanics investigated complaints that several flight attendants experienced watery eyes, scratchy throats, and respiratory symptoms on board a Boeing 747 and a Boeing 777 that had just been overhauled in China. Further investigation found that the personnel in Beijing had used solvents to remove adhesive tape residue on the airplanes' interior panels and carpeting; they mistakenly oversaturated the cabins, causing *off-gassing,* a term for chemical fumes. It was unknown if any passengers fell ill as well. A similar situation occurred with US Airways aircraft serviced by ST Aerospace in Mobile, Alabama; in April 2011 pilots and flight attendants sued the facility. United mechanic David Saucedo points out: "People think when a plane comes back from China it goes through a check flight. Well, it doesn't. Instead passengers get on that plane."

Shortly after I arrived in San Francisco to conduct research for this book, I eased my rental car past the mammoth United Airlines maintenance center on the perimeter of the airport. The airline's site describes it as a "2.9 million square foot facility that is home to more than 5,000 technicians, management, and support personnel." But Saucedo, a twenty-four-year veteran at United, disagrees: "It's a ghost town for us now. We used to have twelve thousand mechanics there and now it's just over two thousand."

Saucedo has been tracking the work done by outside MROs for several years now—and logging quite a few problems. For example, he notes that outside maintenance companies bid on jobs based on a predetermined number of man-hours, so their price is locked in regardless of the actual work required. Imagine asking your local gas station to repair your brakes for a cost you determine up front, regardless of whether you simply need new pads or a complete overhaul. Saucedo explains the second most intensive FAA-mandated airplane maintenance procedure—known as a C-check—has become a cause for concern: "They're finding out that two hundred man-hours isn't enough for a C-check and now it takes two hundred and forty man-hours. But they've already set a determined amount of time. They don't adjust the price, so the vendors cut corners, like not taking out all the panels during inspections. A lot of this stuff seems minor—until something happens."

Connecting the Dots

Of course, occasionally an incident will occur that crystallizes this issue so dramatically it can't be ignored. Like in 2006, when a particularly grisly accident occurred in El Paso, Texas. An outsourced mechanic from Julie's Aircraft Services was trouble-shooting a Continental 737 at the gate, while 114 passengers were settling in on board. He instructed a pilot to start an engine with the cowling open, in direct violation of Boeing's warnings in a manual he had not read. The mechanic was ingested into the engine and suffered a horrific death as traumatized passengers watched.

Unfortunately, as a nation we are not "big-picturing" the severity of the threat of outsourced maintenance. Government white papers emerge. Congressional hearings are held. Investigations evolve from *Consumer Reports*, *USA Today*, PBS. But there's been no drumbeat, no clarion call. Aviation safety is deteriorating in full view, yet the country seems largely unconcerned. "Sooner or later," says FAA whistle-blower Ed Jeszka, "a lot of folks are going to die because of this."

So how are repair stations responding? ARSA has launched the Positive Publicity Campaign Plan (Southwest and Delta have contributed). When I asked Sarah MacLeod why ARSA is seeking positive publicity, she replied, "Negative publicity!" And she cited my own work on this topic: "Mr. McGee, you helped force me into lobbying."

But not all aviation experts are worried by outsourcing. "If the maintenance really was slipshod, we'd know it by now," says Arnold Barnett of MIT. "We should be seeing precursors to accidents." He takes it a step further and questions whether spending more money to provide further protections in an already safe industry is the best use of public resources: "We can even ask, could a billion dollars be used instead on something like cancer research?"

Gordon Bethune is a rare executive. He's an FAA-licensed mechanic, and he uniquely sums up the dangers: "If charlatans can run airlines, they can run repair stations." But Bethune unequivocally states that he is not at all concerned about outside maintenance. He believes the emphasis should be on the total reliability of the operation and says, "I really don't think having an airline uniform and a union will do shit for you with maintenance."

Another factor—albeit misunderstood—is the age of the airplane. A common metaphor is that planes are like cars or even people, and it's all about care and upkeep. In fact, an old industry adage is that there is no such thing as an airplane, only a collection of interchangeable parts all moving in the same direction simultaneously. But like cars and people, older aircraft experience problems specific to their age, particularly with corrosion and metal fatigue. AirSafe.com prominently posts updates on the average age of domestic carriers' fleets, but founder Todd Curtis warns that such information should be taken in context: "It's a half-written story if you just give the age and walk away from it. It's still true that it's about the maintenance and not the age."

Apologists claim there has always been maintenance outsourcing. True, but only to an extent. Before the recent rush, such work traditionally fell into three distinct categories. First, smaller second-tier carriers often relied on outside contractors to perform their work. Second, many carriers have turned to world-class aircraft manufacturers, engine makers, and other suppliers of key parts to service their own products, undoubtedly a sound policy. And third, larger carriers that operate infrequently into "spoke" airports have long contracted with other airlines for "line maintenance" servicing, often on a reciprocal basis.

While some of those smaller airlines may have generated red flags occasionally, none of these trends affected the overall commercial aviation safety picture. But what's occurred over the last decade is much different. I spoke to dozens of frontline airline mechanics and dozens of frontline FAA inspectors based all over America, and it's striking how they sing from the same hymnbooks. As one FAA safety inspector based in the South told me: "We've become complacent. We fall back on 'Our accident record speaks for itself.' "

Of course, all these maintenance issues have generated residual problems such as flight delays and cancellations, as the dreaded "mechanical failure" cannot be classified an "act of God" like so many other flight disruptions, including labor actions.

One longtime employee of a major U.S. carrier says the joke used to be that whenever employees passed a sign that read SAFETY IS #1, they would alter it to read $AFETY IS #1

The Airline Formerly Known as ValuJet

The first wake-up call for the dangers of airline maintenance outsourcing occurred on May 11, 1996, with the crash of ValuJet Flight 592 in the Florida Everglades, killing all 110 aboard. Once again, the NTSB found three parties were to blame: ValuJet; its outside "heavy maintenance" vendor, SabreTech; and the FAA itself. SabreTech had the distinction of being the first U.S. aviation firm to be criminally prosecuted for a fatal airline crash. In fact, a Florida grand jury indicted the company on 110 counts each of manslaughter and third-degree murder. SabreTech pleaded no contest, paid a fine, and soon went out of business. But dozens of other SabreTechs have sprung up in the years since.

In 1997, ValuJet initiated a "reverse merger" by acquiring AirTran Airways and officially changed its name to AirTran, which has since been absorbed into Southwest. (ValuJet, SabreTech, AirTran: maybe the real threat is from corporations that "mid-cap" by imbedding upper-case letters into their names?)

Southwest, of course, has been experiencing maintenance problems of its own. But there is a history of 737s and skin corrosion. On April 28, 1988, Aloha Airlines Flight 243, en route from Hilo to Honolulu, suddenly executed an emergency landing in Maui. Personnel on the ground were shocked to see that an eighteen-foot section of the fuselage skin over the entire forward cabin of the Boeing 737 was missing, and several rows of passengers were effectively sitting in a convertible. What the NTSB termed an "explosive decompression and structural failure" at twenty-four thousand feet meant a small crack near the cabin entrance door had suddenly ruptured, causing the cabin skin and structure to peel away, depressurizing the airplane and disabling one of the two engines. A flight attendant in the forward section was sucked out over the Pacific Ocean and her body was never recovered. The NTSB later noted the nineteen-year-old island-hopping plane had accumulated nearly ninety thousand pressurized takeoff-and-landing flight cycles, the second-highest total of all 737s worldwide (behind an Aloha 737 sister ship).

Twenty years later, in 2008, a congressional hearing brought to light that Southwest operated at least 117 aircraft without performing critical required inspections on hydraulic systems and cracks in the fuselage, a situation one congressman called "the most egregious lapse of safety I have seen in twenty-three years." Scott J. Bloch, special counsel for the U.S. Office of Special Counsel, testified that two FAA inspectors who became whistle-blowers disclosed that the FAA principal maintenance inspector for Southwest knowingly permitted aircraft to operate approximately 1,400 flights with those fuselage cracks, putting passengers at risk.

I asked Gary Kelly, the CEO of Southwest, about this and he responded that the well-publicized rupture in the fuselage of a Boeing 737 in April 2011 was a manufacturing defect and not a maintenance issue. He added, "The word *outsourcing* is irrelevant. Southwest Airlines is completely, one-hundred-percent responsible for the maintenance of our aircraft, whether we use employees or contract maintenance and repair organizations."

Mary Schiavo, the DOT's inspector general from 1990 to 1996, made national headlines when she fought for the shutdown of ValuJet and assailed the FAA for its lack of oversight. Ten years after leaving office, she told me that not only had the maintenance outsourcing situation not improved, but in fact "the playing field is tilting the other way."

North by Northworst

Like ValuJet, Northwest Airlines doesn't exist anymore; in January 2010 it was completely absorbed by Delta. However, the lessons of Northwest's maintenance program will persist for years to come, and have not been lost on airline executives, labor negotiators, or federal regulators. Basically, the entire industry watched a major airline break its mechanics union and then proceed to openly operate aircraft in what was unquestionably a much more dangerous manner—while the FAA watched and did next to nothing. All of us live with this legacy every time we board a domestic flight.

The nation's fourth-largest airline, Northwest, had a fleet of 369 airplanes when the Aircraft Mechanics Fraternal Association went on strike in 2005. The in-house maintenance staff was reduced from a reported figure of 3,600 to just 900 overnight, and the FAA publicly stated it would increase oversight of the Minneapolis-based airline. So here is the total amount of maintenance fines levied by the FAA against Northwest during the strike: $0.00. Zero dollars, zero cents. A major U.S. carrier operating worldwide reportedly lost 75 percent of its maintenance force and yet the FAA asserted that no corners were cut. Eventually management won, of course, and the strike was settled after 444 days (the exact number of days American hostages were held in Iran).

Not surprisingly, all evidence clearly suggests there were multiple violations, and news reports documented dozens of issues. In early 2006, a Northwest aircraft en route to Japan suffered a hydraulic problem and was forced to execute an emergency landing in San Francisco, an event an official termed a "normal emergency landing." That launched a satiric blog titled *Weekly Normal Emergency Landing,* though that satire was rendered impotent when Northwest suffered *two* emergency landings in as many days. But even that paled beside the Northwest plane that executed two *back-to-back* emergency landings in Springfield, Missouri, because of smoke warnings, both of which were false.

A former Northwest mechanic who was stationed for a time at the airline's hub in MSP recalls two DC-10s that returned from heavy maintenance service at SASCO in Singapore needing critical repairs before reentering service: "The pilots would go pick up a plane, and before they landed the [maintenance] log book was filled."

According to former FAA employees, a pattern emerged. Allegations were officially denied; inspectors were browbeaten, harassed, and transferred; lives were ruined; and potentially life-threatening conditions were ignored. One FAA inspector claimed almost *five hundred reports* on Northwest in the first eleven days of the strike were never entered into the FAA's database. He also asserted between 58 percent and 90 percent of those reports noted "defects" in Northwest's operations; an airline's defect rate is usually below 5 percent, and a defect rate of just 9 percent automatically triggers an FAA audit.

An FAA inspector named Mark Lund became a whistleblower after his claims about the agency's lax treatment of

Northwest went unaddressed. When he received official satisfaction, it was a cold dish indeed: in 2010, a report from the U.S. Office of Special Counsel substantiated Lund's allegations that an FAA office "failed to provide effective oversight of Northwest's [FAA Airworthiness Directives, or AD] process, resulting in the carrier's continued systemic AD non-compliance." The report acknowledged that Northwest had not complied with ADs for more than a decade and the status of more than one thousand outstanding ADs was unknown. The FAA responded by forming an Internal Assistance Capability review team. But by 2010, of course, many of Northwest's mechanics had moved on to other jobs, FAA inspectors had seen their careers threatened, and the airline itself was absorbed into Delta. And Richard Anderson, the former Northwest CEO, had become Delta's CEO.

Since poring over FAA records can be a dead end, a few years back I turned to NASA, which maintains a valuable resource called the Aviation Safety Reporting System. ASRS generates feedback from industry professionals and publishes their warnings about specific problem areas. NASA redacts the names of individuals, since it wants critical and even life-threatening information to be shared freely. ASRS is a necessary and vital communication channel, though I don't believe airline names should be omitted.

If a posting to ASRS is serious enough, NASA will issue an Alert Bulletin, and that's what it did in 2005 with a "critical" problem on an unnamed domestic airline, affecting dozens of its aircraft. The bulletin read: "Technicians reported finding broken, loose, and missing wheel tie bolts on wheels built up by a contract-maintenance facility." A contract-maintenance facility, eh? Hmmm. Although the airline's name wasn't provided, the types of affected aircraft were, so I cross-referenced the airplane models with the fleets operated by every U.S. carrier—and all evidence pointed to Northwest. I contacted Northwest and inquired of a public relations rep if the wheel tie bolt problem had been addressed. Then I waited. He soon got back and assured me the problem had been fixed, thereby confirming the identity of the airline NASA had refused to disclose.

Why was this critical and potentially life-threatening problem brought to light through a voluntary reporting system operated by NASA and not through the FAA, the agency in charge of overseeing Northwest's maintenance? Why were no fines levied? And in an age of supposed government transparency, why wasn't the FAA publicly reporting on such a critical safety issue? In fact, if the problem was so widespread and so threatening to passenger safety, why wasn't the fleet grounded until all wheel tie bolts were repaired?

Like all airlines, Northwest knew that farming out maintenance and aircraft handling to lowballing contractors was risky. I obtained a copy of an internal Northwest email from a senior executive dated April 2006, warning airport managers that outsourced employees could create safety concerns: "We are certain of at least one thing: changes and inexperienced employees running equipment. . . . This is one of those failures we can actually prepare for." There you have it. A senior airline exec acknowledging that outsourcing leads to "failure."

Although the lessons of ValuJet and Air Midwest were lost on the airlines, the lessons of the Northwest strike have been taken to heart.

Back in the U.S.A.

If there is *any* reason to believe major carriers might once again service their fleets on U.S. soil, ironically enough it might derive from United, a company in the forefront of shipping jobs overseas. In 2009 the Teamsters approached United chairman Glenn Tilton and asked to bid on some outsourced jobs and to perform the work at that underutilized maintenance base in San Francisco. A small step, yes, but a tentative and still shaky alliance is being forged. David Bourne says, "Now we're insourcing and bringing work back in [to the United States]. We know we have to be competitive." Does it inspire a little optimism? "Absolutely! But you can't get it done overnight."

After so much investigating of maintenance outsourcing, I decided it was time to examine the lone exception. I planned a two-day visit to the American Airlines Maintenance and Engineering Base in Tulsa, the largest private employer in Oklahoma. I wanted to meet what I desperately hoped was not an endangered species: U.S. airline mechanics.

As the last bastion, American actually faces two herculean tasks: the first is continuing to service its own planes at a competitive cost disadvantage, and the second is attracting new business through insourcing.

I arrived at "the largest commercial aircraft maintenance facility in the world" as a joint guest of both American and the Transport Workers Union, and spent time with both management and labor. I walked several miles to tour the 330-acre plant, which employs 6,700 workers and was established in 1946 when the airline bought a B-24 complex at a postwar surplus price. Over the years Tulsa has expanded to accommodate jets, widebodies, and the aforementioned "Growth Plan" in the 1980s.

Simply put, it was an impressive tour. The facility is a small village and even includes manufacturing and machine shops ("every bushing from scratch"), costly investments smaller airlines could never make. But another costly investment is the workforce, which is highly trained and overwhelmingly licensed by the FAA. "There is a less expensive way to go," says Bill Collins, vice president of base maintenance for American. "We pay a bit of a premium to have all our mechanics certified."

Yet last year the FAA proposed a $24.2 million civil penalty against American for failing to inspect wire bundles in the wheel wells of MD-80 aircraft, the highest fine ever proposed in FAA history. This from the same FAA that levied *no* maintenance fines against Northwest when nearly its entire workforce walked off the job. How could this be?

There's a sticky question hanging, one that's been whispered within the industry for years now, and I finally asked it in Tulsa: Is American being unfairly targeted by the FAA because unlike all its domestic competitors, it operates right under the proverbial noses of federal inspectors? After all, the FAA has an office right outside the gates on North Memorial Drive and averages two visits per week, and sometimes as many as six inspectors are on the premises at once. So is American an all-too-easy target in Tulsa, unlike those in China or El Salvador? Brad Mueller of the TWU provided an intriguing response by suggesting that FAA inspectors overseeing work in Asia or Latin America be asked: "Do they have dedicated parking spots out front there like they do with us?" The answer is obvious.

American's relations with the FAA are "cordial," according to Collins, but he diplomatically added, "It's not something that gets talked about with senior management." However, he did suggest he has it easier than his counterparts at other carriers: "I can sleep at night because everything has been thoroughly vetted."

Collins said, "We believe it's a competitive advantage. The risk mitigation factor is considerable. . . . It's about having an accident or not having an accident. By orders of magnitude, the premium is justified in that way."

That seems like inarguable logic. Yet many other airline executives, board members, analysts, and investors clearly don't justify such a premium. Now all eyes are on American's bankruptcy proceedings.

Reagan Revolution Home to Roost

The airline industry has woven the defense of outsourcing into a defense of capitalism itself. Airline execs continually resurrect the dour Calvin Coolidge to intone, "The chief business of the American people is business." As for the wisdom of employing unsupervised minimum-wage workers to maintain wide-body commercial jets when they can't read the repair manuals, free-market advocates characterize those who beg to differ as being just to the left of Fidel Castro.

Someone with the battle scars to prove it is former DOT inspector general Mary Schiavo; a few years ago she told me, "Back in the 1990s I was saying we need to reregulate certain aspects of aviation—if you said that, it was like you were the devil yourself. Government needs to do for the citizens the things they can't do for themselves. Some said airlines will regulate themselves—it was a great myth. The airlines do not and cannot do it."

Instead, we have an FAA that has called airlines its "customers," a phenomenon that's occurred elsewhere, such as with the National Highway Traffic Safety Administration and Toyota. It all comes back to cost. Consider this chilling statement from NTSB chairman Deborah Hersman: "You can't get through a cost-benefit analysis without the numbers. You have to wait for people to be killed."

I asked Secretary LaHood, at what point does maintenance outsourcing become a government issue? He responded, "Oh, I think it becomes a government issue when the FAA bill gets considered and foreign repair stations are included and that's when we begin to weigh in and see how that plays out."

Unfortunately, Congress won't step in soon. As recently as March 2010, Senator Claire McCaskill, the Missouri Democrat, was pushing for an amendment that would prohibit work at noncertificated repair stations and increase FAA inspections and drug and alcohol testing. But a source on Capitol Hill told me the amendment was dead, primarily because Senator Jay Rockefeller of West Virginia would actively oppose the amendment over concerns it "would violate the safety treaty language the FAA signed with foreign nations." For the record, ARSA reports that repair stations in West Virginia employ 1,448 workers, with a total economic impact on the state of $157.1 million. (Neither McCaskill nor Rockefeller responded to repeated requests for interviews.)

The policies that now threaten our nation's aviation safety record can be traced to 1981, when President Ronald Reagan came into office warning "government is not the solution to our problem; government *is* the problem." But today the real and present danger threatening airline safety is weak enforcement, thanks to nearly three decades of the Reagan "smaller government" mantra. Reagan's infamous quip that the nine most terrifying words in the English language are, "I'm from the government and I'm here to help" should be considered in the context of an FAA that treats airlines—not the flying public—as its customers.

Has the federal government implemented nonsensical regulations? No question. Has it wasted tax dollars? Absolutely. Would bureaucracy impede many of the industry's brightest innovators if the government completely ran the airlines? Certainly.

But demonizing regulators and placing all trust in the free market is a dangerous proposition as well. When tragedy strikes the airline industry, apologists claim that executives don't want such accidents to occur. Of course, one would hope as much. But airline executives also should not be the last line of defense in *preventing* such accidents. Based on my experiences working in and investigating the airline industry, I can personally testify that there truly are airline executives who cut corners on safety. That's irrefutable fact.

8

Unsafe at Any Altitude? Facing Unprecedented Dangers

Courage is the price that life exacts for granting peace.
—*Amelia Earhart, pilot*

LET'S BE VERY CLEAR: COMMERCIAL aviation is the safest form of transportation available. What's more, it is particularly safe in the United States—though the statistics are quite different when regional carriers are broken out separately. But this is an industry that suffered no domestic fatalities in 2011.

The most dangerous year for airlines was 1929—fifty-one fatal crashes—but the record has steadily improved for more than eighty years. According to Boeing, fatal accidents occurred about once every 200,000 flights in the 1950s and 1960s; worldwide, the record is more than ten times better today, with fatal accidents occurring less than once every 2 million flights. The Insurance Institute of America states the lifetime odds of dying in a car accident are 303 to 1 now, while dying in an air or space accident are 7,032 to 1.

"Aviation is in a league of its own when it comes to safety," says Deborah Hersman, chairman of the National Transportation Safety Board. She notes: "More than one out of five trucks are unsafe when you drive alongside them to visit grandma. If we had that kind of maintenance issue with airplanes, no one would fly. We tolerate deficiencies in other modes of transportation regularly. The bar is set differently with aviation."

The true experts on aviation safety speak in their own language. "Safety is sort of like beauty," explains Todd Curtis, founder of AirSafe.com; "it's in the eye of the beholder." His résumé is impeccable—Princeton, MIT, the Air Force, Boeing—and his site is a trove of airline safety statistics, diced and sliced by time lines, airlines, aircraft. But every mathematical equation and nugget of information comes with a caveat, because wading into safety is all about lies, damn lies, and statistics.

Airline safety often comes down to body counts. As Curtis says, "When it comes to the amount of ink spilled, what really drives the issue is if passengers are killed." Just as homicides are not all treated equally by police, aviation accidents are not treated equally by regulators, legislators, and journalists. High-profile accidents often generate more resources for investigations and remedies.

Identifying the most important risks isn't easy. The list includes everything from bird strikes to volcanic ash to laptop batteries stored in overhead bins. New threats arise continually—airlines are removing life rafts from aircraft and oxygen masks from onboard lavatories (or is that a good thing, since it reduces the terrorist threat?).

A Risky Business

The airline gods were busy on April Fool's Day 2011. I was in snowy Boston, meeting with two well-known airline safety experts, MIT professor Arnold Barnett and former NTSB member John Goglia. Meanwhile, on the floor of Congress, Republican lawmakers were pushing through an FAA reauthorization bill that called for slashing $4 billion from the agency's budget—

cuts that experts said would most certainly lead to inspector layoffs. On the same day, Peggy Gilligan, the FAA's associate administrator for aviation safety, was speaking at the Aeronautical Repair Station Association's Annual Repair Symposium. Simultaneously, a five-foot hole was suddenly opening in the roof of a Southwest Airlines Boeing 737 en route from Phoenix to Sacramento. Clearly the theme of the day was risk.

In the end, safety is all about risk and reward. And the FAA—like all government agencies—must conduct analyses of rules and directives based on lives saved and lives lost. It's messy and controversial work. Little wonder the mayhem depicted in the novel *Fight Club* is told by our unnamed protagonist, a Product Recall Specialist analyzing defective automobiles on a cost-benefit basis in the wake of fatal car accidents.

"Change is risk," says Tom Brantley, president of Professional Aviation Safety Specialists, a union representing FAA employees. He notes that the FAA has changed how it measures accidents, by creating separate silos for large and small airlines, and therefore "it's not apples to apples." He also cites a combination of troubling factors, including an aging fleet of commercial planes in the United States, less experienced cockpit crews, and a loss of "tribal knowledge" passed on through generations of mechanics.

At MIT, Barnett says, "A lot of supposed risks turn out to be statistical mirages." He adds that new technological data mining techniques have gotten so sophisticated they can ask logical questions about risks. "So there is reason to be hopeful," he says. "You can see anomalies that did not result in accidents. So you can anticipate problems."

But in an era of fewer accidents, measuring risk has become harder. This is particularly true because old gremlins—such as wind shear, lightning strikes, and some types of pilot error—have been largely addressed, so that today causes can be more complex.[1] Curtis asserts that had the TWA Flight 800 explosion off the coast of Long Island in 1996 occurred on another Boeing 747—say a cargo model in a rural area—the NTSB would never have spent four years on the investigation.

Dangerous Ground: Runway Incursions/Excursions

The FAA defines a runway incursion as "any occurrence at an airport involving an aircraft, vehicle, person, or object on the ground that creates a collision hazard or results in a loss of separation with an aircraft taking off, attempting to take off, landing, or attempting to land." Such events are further classified by a severity index, with the most dangerous occurrences labeled Category A, those in which "extreme action [is taken] to narrowly avoid a collision, or in which a collision occurs."

The NTSB has been vocal about this for years. Among the "Most Wanted" actions it is seeking from the FAA are the following: flight crews should be given immediate warnings of probable collisions; each runway crossing requires specific air traffic control clearance; airlines should install cockpit moving map displays or an automatic system to alert pilots when the wrong taxiway or runway is broached; and a "landing distance assessment" should be devised for every landing.

The NTSB's Jeff Marcus explains that the FAA collects a lot of statistics on near-misses, and there are differing severities that greatly affect the significance of these findings; the worst are incidents that fall into the A to D categories, in which aircraft are just a few feet apart. He contends that the real issue is not the total number of events, but how these events relate to traffic statistics. Overall, he contends that the FAA has made "real progress" in addressing these threats.

In its Annual Runway Safety Report for 2010, the FAA stated

there were 951 runway incursions in 2009, down from 1,009 in 2008; furthermore, there were 12 "serious" incursions in 2009, down from 25 the year before. The report concluded: "When serious runway incursions drop by 50 percent over the previous year, you know you're doing something right."

But the news is not all bright. Whistle-blower Richard Wyeroski's career at the FAA entered a downward spiral after he began investigating not one but two dangerous Category A incursions that occurred within the same week at Long Island MacArthur Airport in Islip. Apparently the real problem was that both events involved major airlines: first Delta, then Southwest.

On March 4, 2000, a Delta jet was cleared for takeoff on Runway 24, even though a Cessna 172 had just landed and was still taxiing on 24; Wyeroski estimates the Boeing 737 cleared the roof of the Cessna by less than fifty feet. He recalls that the pilots in the Cessna were so traumatized they couldn't exit the airplane on their own. That same week the MacArthur tower also cleared a Southwest 737 for takeoff, right in front of a small aircraft that had been cleared to land on the same runway (luckily the pilot of the smaller plane took evasive action and aborted the landing).

Wyeroski maintains that reporting these two events caused his dismissal. After being fired by the FAA, he later filed a statement with the Disclosure Unit of the U.S. Office of Special Counsel in which he claimed he had been told by an FAA supervisor that Wyeroski's boss was "really pissed" over his reporting of the incursions.

A Fatigued Industry

The NTSB's original "Most Wanted" list included improvements in combating human fatigue, but since then, not much has changed. Indeed, the NTSB's latest "Most Wanted" list still includes Human Fatigue in the Aviation Industry. Specifically, the NTSB suggests setting working-hour limits for flight crews, aviation mechanics, and air traffic controllers based on fatigue research, circadian rhythms, and sleep and rest requirements. It also recommends developing guidance for operators to establish fatigue management systems, including a methodology that will continually assess the effectiveness of these systems.

According to the National Air Traffic Controllers Association, "over 14 accidents resulting in 263 fatalities had fatigue as a causal or contributing factor" since 1993. Then in the spring of 2011, the issue became front-page news after a controller fell asleep at Washington National Airport; in the weeks that followed, similar incidents occurred throughout the country—Lubbock, Reno, Seattle.

Congress immediately spoke out, and in a Senate hearing Senator Jay Rockefeller stated: "We shouldn't tarnish the whole profession based on the poor judgment of a few. . . . We can't allow recent questions about the safety of the FAA to permeate air travel." Brantley says the sleeping controllers were "embarrassing" for the FAA. After all, the agency is responsible for ensuring safety in American aviation, yet air traffic controllers are FAA employees.

I spoke to then administrator Babbitt about this and even admitted to my own snoozing back at Pan Am Shuttle. Here's what he had to say: "We're rewriting the rule for pilots and we're taking a similar approach. We pay people to come to work rested and ready." But he also encourages aviation professionals to speak up about this topic: "In this day and age we encourage people to self-report. Tell us . . . don't sit there and be lost in space."

As with many issues related to airlines, there is a communications disconnect. When news reports of sleeping controllers first broke, angry politicians and pundits demanded explanations.

Meanwhile, industry professionals shrugged—who didn't know about napping on the job? FAA whistle-blower Ed Jeszka provides perspective: "Do you know how long that has been going on? Ask pilots who fly night flights. Very often they are handed off [from one air traffic control facility to another] and no one responds on the radio."

In fact, the dirty little secret of aviation—and perhaps many other critical 24/7 professions—is that napping at work has always existed. It certainly did in Operations Control when I was at the Pan Am Shuttle. A pilot confirms it was standard procedure for cockpit crews on red-eye flights.

This is in no way to excuse what can be dangerous behavior. Experts point out there obviously are two huge caveats here: (1) critical personnel still must be available to respond immediately; and (2) when two or more are on duty, they need to take shifts, just as the military rotates guard duty in hostile environments. But now aviation professionals are attempting to better understand circadian rhythms and find solutions.

Marcus says the NTSB "butted heads for a long time" with the FAA and the National Air Traffic Controllers Association over controller fatigue—and this was before a spate of "sleeping controller" stories dominated headlines. However, he is optimistic new FAA staffing will address this problem.

Foreign Carriers: Who's in Charge?

In the 1990s many aviation professionals began asking who oversees foreign airlines operating in and out of U.S. airspace. The answer, of course, was and is the FAA, but the issue garnered more attention after the crash of Avianca Flight 52 one foggy night in 1990, when a Boeing 707 simply ran out of fuel and fell from the sky over Cove Neck, New York, fifteen miles short of John F. Kennedy International Airport. As a licensed dispatcher, I was particularly intrigued by the NTSB's summation: "Contributing to the accident was the flight crew's failure to use an airline operational control dispatch system to assist them during the international flight into a high-density airport in poor weather." Aviation professionals were stunned that seventy-three people were killed because a passenger airline did not have a dispatcher on duty (a requirement for U.S. carriers since 1938), and its Spanish-speaking crew members were unable to convey to controllers the seriousness of their fuel depletion, requesting a "priority" rather than an "emergency" landing.

Back in 1994 I wrote a lengthy article for *Air Transport World* titled "Two Sets of Rules," which detailed how foreign airlines adhere not to Part 121 of the Federal Aviation Regulations but to a much more lenient interpretation, Part 129. I also reported that the FAA maintained a "blacklist" of countries banned from allowing their aircraft to operate here, but the list was not made public. Anthony Broderick, then the FAA's associate administrator for regulation and certification, stated: "If we release all of our documents to the public, that's not appropriate. And it can have a chilling effect on our relationship with that other government."

Thankfully, there is more transparency today. Under the International Aviation Safety Assessments program, the FAA states it has established two ratings for countries at the time of assessment: Category 1, the nation complies with International Civil Aviation Organization standards, or Category 2, it does not. (ICAO, which is chartered by the United Nations, establishes worldwide safety standards for airlines but has no enforcement authority.) The 2011 list consists of 103 countries assessed, of which 22 do not meet ICAO benchmarks; this list overwhelmingly consists of developing nations in Latin America, Africa, and Eastern Europe, though it does include Indonesia, Israel, and the Philippines. In the case of Israel, for example, the FAA cited "areas of concern" with the country's civil aviation authority following an assessment in 2008, but it hasn't prevented Israeli aircraft from operating here under "heightened surveillance"; however, the FAA says "Israeli air carriers will not be allowed to establish new service to the United States" until the Category 1 rating is regained.

Unfortunately, frontline FAA inspectors say one problem remains: the State Department has more say over such matters than FAA experts. What we're left with, according to experts, is an FAA that doesn't have the final say on which airlines can fly into the United States.

Pilot Proficiency Still Wanting

Among those NTSB "Most Wanted" improvements for aviation, the first is improving oversight of pilot proficiency. Specifically, the board recommends that the FAA evaluate prior flight check failures for pilot applicants before hiring and provide training and additional oversight that considers full performance histories for flight crew members demonstrating performance deficiencies.

There are two aspects to the pilot proficiency problem. One is ensuring that an airline hires qualified pilots. The other is dealing with inadequate record keeping that prevents airlines from effectively evaluating performance. The FAA and NTSB have disagreed about citing pilots with repeated histories of failed flight checks, such as the captain of Colgan Air Flight 3407.[2] The NTSB's proposal is to identify pilots with remedial problems and provide them with remedial training.

Hopefully the Airline Safety and Pilot Training Improvement Act, introduced by Congressman Jerry Costello in 2009, will do just that—improve safety. Costello told me: "I'm a strong believer in holding the FAA's feet to the fire. We need pressure and transparency."

Unseen Dangers: Bogus Airplane Parts

For several years now, FAA inspectors and airline mechanics have been telling me about black-market parts making their way onto U.S.-registered commercial aircraft. Even determining the scope of the problem is difficult, since there are fewer FAA inspectors and airline mechanics present when heavy maintenance work is performed. A former Northwest mechanic says it was common for aircraft returning from service overseas to be outfitted with defective parts, and recalls a supervisor's warning: "We've had a rash of bad parts coming out of stock." The mechanic adds, "Forget it—by the time you find out about it, it's too late."

Chris Moore of the Teamsters Aviation Mechanics Coalition sums it up: "Here's the credible threat. The airlines in-country [in the United States] do a really good job of controlling their inventory. But in a Third World country, who's watching? Are you going to wait for a part to be shipped in? And who's overseeing it?" I also received an anonymous communication indicating that incidents with failed parts are increasing tenfold, with both new and refurbished components. John Goglia says, "For a while the FAA put a lot of manpower on this. Now it's creeping back in."

Another mechanic for a major carrier told me of a related but equally disturbing problem: "We do see improperly assembled parts from outside repair facilities. We never saw that in the past.

We've never seen it as bad as it is now. We have more 'bad from stock' parts than we've ever had." He explained how stocked parts are often rebuilt parts, and these days such work is done by outsourced shops: "Reverse actuators are rebuilt stock. To replace one we may have to go through our entire stock to find one that has been rebuilt properly. If it's not rebuilt properly, it can lead to smoke in the cabin." And that, of course, can lead to a catastrophe. And all roads lead back to the FAA.

The FAA: Doing Right by Its Customers

The Federal Aviation Act of 1958—which created the Federal Aviation Agency, the predecessor organization of the current Federal Aviation Administration—elevates safety in its first declaration: "The assignment and maintenance of safety as the highest priority in air commerce." But despite the gravity of this credo, whistle-blower Gabe Bruno declares, "It's no longer relevant to the FAA."

That's a strong charge, and quite frankly I was skeptical when I first began speaking to former FAA employees and other aviation professionals who render such harsh assessments. Has their anger clouded their vision about the government agency that oversaw nearly twenty-nine thousand commercial flights per day in 2010 without a single fatality? Yet eventually I developed tremendous respect for Bruno and the other whistle-blowers who have taken tremendous personal risks to expose what's wrong with the FAA. And I came to believe their charges.

There's no denying the FAA is one of the broadest targets in the federal government. It is assailed from all sides—Congress, the industry, the media, consumers. And the criticisms usually take several forms:

- FAA is beholden to the airlines
- FAA is slow-moving and excessively bureaucratic
- FAA is a schizophrenic collection of fiefdoms

Opposing Regulation Ad Hominem

It's a given that nearly all corporations will instinctively oppose, knee-jerk fashion, *any* form of regulation, regardless of the benefit to consumers, the national welfare, or the health of the planet. David Bollier, Ralph Nader's biographer, detailed how the auto industry responded to safety legislation: "Speaking for many of his Detroit colleagues, Henry Ford II complained that the new auto safety standards were 'unreasonable, arbitrary, and technically unfeasible . . . if we can't meet them when they are published we'll have to close down.' But in 1977, an older and wiser Henry Ford conceded on *Meet the Press*, 'We wouldn't have the kinds of safety built into automobiles that we have had unless there had been a federal law.' "

In the war for strengthening airline safety standards, many experts contend the airlines are winning—hands down. In 2008, James P. Hoffa stated that United Airlines "does maintenance in Beijing with 2,200 mechanics, and only five of them are FAA-certified. That, to me, is a threat." I asked Hoffa about FAA oversight, and he said, "Big industry does not want food inspectors, mine inspectors, bridge inspectors. Big business does not want to be regulated." As Bill Voss of Flight Safety Foundation says, "The Air Transport Association [now Airlines for America] needs to outgrow the mentality that regulations equal bad."

Eighteen years ago Nader cowrote *Collision Course: The Truth About Airline Safety*, which asserted that with airline safety "the level of risk appears to be growing." In 2011 I asked Nader about his 1994 book and he responded, "It's the same thing now. Nothing has changed. It doesn't matter if it's a Republican or Democratic administration. They're constantly behind the eight ball and they're constantly allowing the airlines to self-regulate." He added, "It's a charmed life. It's amazing there haven't been more crashes."

Nader argues that the FAA has devolved into "basically a bureaucratic secretariat." But his view of the agency's safety oversight is particularly stinging: "They're lucky. In that when something goes wrong there's instant accountability. It's called the pilot." He articulates a view that many aviation professionals dare speak of only in whispers: "pilot error" covers any manner of sins in the wake of serious accidents, particularly when cockpit crews don't survive.

As for maintenance outsourcing, the key issue is FAA oversight. It's a long-standing problem, dating back to the agency's earliest days, when its "dual mandate" called for both regulation and promotion of the airline industry. Eventually Congress eliminated the *promotion* part in 1996, but critics say the mind-set remains the same.

Consider the watershed moment in 2003, when FAA administrator Marion Blakey, an appointee of President George W. Bush, introduced the FAA's "Customer Service Initiative." In a speech before industry players that February, Blakey stated: "And, we want to know from our customers if we're not being consistent. We're going to let them know that they have the right to ask for review on any inspector's decision on any call that's made in the certification process." Not lost on anyone—least of all frontline FAA inspectors—was the odd use of the word *customer* to describe the corporations a government regulatory agency was charged with overseeing.

It's a topic that came up more than a dozen times when I spoke to enraged FAA employees. As one put it: "Our customers are the American people, not the airlines." It's duly noted that shortly after President Obama took office, the FAA Mission Statement was altered and "customers" became "stakeholders." And once Randy Babbitt became administrator, the FAA issued a public pronouncement that the definition of CSI had changed, from Customer Service Initiative to Consistency and Standardization Initiative. But critics roll their eyes at the semantics and instead note that the endemic problems stem from the FAA's culture, not its written mandates.

Sometimes a headline can capture it all, as in this gem from the *Washington Post* in 2002: "Improper Use of Tape to Fix Wings May Lead to FAA Fine for United." The keeper in that phrase, of course, is *may*. At issue were three Boeing 727s with holes in their ground spoiler panels—critical control surfaces on the wings—that United elected to repair with what is widely known in the business as "speed tape." Since the three planes were improperly flown on a total of 193 flights, the total fine levied by the FAA should have been $2,123,000, but the initial penalty was immediately knocked down by 62 percent. The *Post* quoted a letter from the FAA: "Having considered all the circumstances in this case, we would be willing to accept an offer in compromise in the amount of $805,000 in full settlement of this matter." Brantley estimates such final judgments are usually reduced to about 10 percent of the total: "They kind of abuse the concept of it not being punitive."

Karen Eckert, who maintains that the Colgan Air crash that killed her sister and forty-nine others was not an accident at all, since the contributing factors were both "controllable and predictable," says, "It's profit first, safety second." She says flying is "a horror show," because the FAA has gotten too cozy with the airlines it oversees. Although she asserts this is an understandable manifestation of human nature, she suggests that teams of inspectors rotate to avoid such issues.

The coziness charge is nothing new. As one FAA inspector

told me, "A lot of these guys came from the airlines and they still think like they're airline guys. It's not that they'll cover up everything. But they are gonna soft-sell certain things."

However, the NTSB's Tom Haueter rejects the coziness theory: "It goes back to laziness. It's more about not doing the job. To me, it's not cozy, it's just lazy." So is the FAA divorced from reality? "You need boots on the ground," says Haueter. "The FAA philosophy is 'We're working with an honest man.' I think they should not trust anyone. There are competitive pressures in this industry." He says this FAA mind-set even extends to treatment of aircraft manufacturers: "The 787 is essentially being certified by Boeing. Where's the FAA oversight?"

Critics also claim that the FAA's self-disclosure programs allow airlines to have it both ways: address apparent concerns yet avoid any penalties. Brantley says, "The idea of self-reporting is you get immunity. If you think you're going to get caught anyway, it's a huge way to protect yourself. That's human nature."

The FAA has earned a nickname that is well-known throughout aviation: the Tombstone Agency, meaning the FAA responds only *after* a fatal accident, not before. I brought it into the open with Randy Babbitt himself prior to his abrupt resignation, and asked if he thought it was fair. He responded: "You're right. There has been the charge that it's a tombstone agency. What we have done is increase voluntary reporting, so people can tell us. And I think it's taken hold a little bit." He also pointed out that "when I came here I didn't appreciate how complex the FAA is. I thought I knew a fair amount and I didn't."

I asked Secretary LaHood if he worries about his workforce getting too close to the industries they oversee, and he responded: "Not at all. Not at all. We have people that have worked here for years and they're very professional people. They don't get too close to the industry. As a matter of fact, they really go out of their way to make sure they're doing things for the right reason, for the people who are riding these different forms of transportation."

Controlling the Fiefdoms

One of the toughest questions is whether any FAA administrator can oversee such a large, far-flung organization. In Babbitt's view, "You need to get on a bully pulpit. The only way to improve is better collection and analysis of the data. So you can identify risk everywhere you see it."

Voss says the system is designed so the FAA administrator does not have "a whole lot" of effect on the organization's culture. He explains: "With all these huge stinking problems, there's no one who can sit in that seat and steer it. And Congress really enjoys that it's not succeeding. It's cynical, but there's no investment in changing. The FAA is the donkey you stick the tail on. It's an agency that is designed to take blame and risk." He adds, "You do have a culture that is extremely change-resistant."

Others speak about the lack of uniformity in interpreting and enforcing regulations from coast to coast. Industry veterans note that airlines frustrated by a given Flight Standards District Office will consult with another office. Even NTSB employees speak of "FSDO shopping," whereby airlines that don't like mandates issued by a given FSDO simply find another FSDO that interprets the regulations differently.[3]

John Goglia, the only FAA-licensed airframe and powerplant mechanic to have served as a board member on the NTSB, is characteristically blunt: "It's not budget. That would help but it's not enough. It's culture and it's how to do the job." I asked him for an update: "Nothing has changed. If anything, it's worse. The FAA can't do the job. . . . They're jerking each other off. The airlines love it when the FAA doesn't do its job." Occasionally there are cases when this becomes apparent.

Alaska: An Airline Without an FAA

After Earlene Shaw lost her husband, Don, in the crash of Alaska Airlines Flight 261 in January 2000, she and other next of kin began an intense journey to uncover the truth. The Shaws, who live in Enumclaw, Washington, and have relatives in Alaska, had been frequent flyers on the Seattle-based carrier, with about 185,000 miles in their account. But when the systemic causes of the tragedy came to light, Earlene was stunned by Alaska's actions, and the FAA's as well. She sums it up: "I must say that to this day the FAA shares a great deal of responsibility for that accident—along with Alaska Airlines and its maintenance. There's a lot of anger and disappointment that a government agency did such a poor job."

Flight 261 was a particularly harrowing accident. Those on board were terrorized in their final moments after the crew lost control of the McDonnell Douglas MD-83 and the plane dove vertically into the Pacific Ocean off the California coast, killing all eighty-eight on board. The cause? Insufficient lubrication of the jackscrew that held together the critical horizontal stabilizer assembly, thereby preventing the pilots from controlling the aircraft's ability to ascend and descend. But the unusually lengthy NTSB accident report includes detailed discussions of a range of Alaska and FAA deficiencies. As one former FAA inspector told me: "That jackscrew was the last piece of that accident."

The real problems had begun long before that night. Two months after the accident, the *Seattle Times* published a letter to the airline's CEO from sixty-four Alaska mechanics noting "our pleadings have gone unheeded." The letter stated that Robert Falla, manager of base maintenance in Seattle, "pressured, threatened, and intimidated" mechanics: "He has directed us to do things specifically contradicting [Federal Aviation Regulations], not the least of which is his persistent demand that we put unserviceable parts back on the aircraft. . . ." By March 2000 the FBI had opened a criminal probe into Alaska's maintenance program, since falsification of federal records is a crime (the case faded away quietly under Attorney General John Ashcroft's watch).

Prior to assuming the presidency of the Aircraft Mechanics Fraternal Association, Louie Key spent thirty-two years as a mechanic for Alaska—including the dark days before and after the fatal crash of Flight 261. He describes Alaska's corporate culture at that time as obsessively focused on Wall Street's industry measurement of Cost per Available Seat Mile: "It was absolutely frustrating. Everything was about cutting costs. It was 'We've got to get CASM down.' Unfortunately, it had consequences."

As part of the accident investigation team, Key was on the recovery vessel that retrieved the jackscrew, which the NTSB determined had failed in-flight due to "excessive wear" caused by a lack of lubrication. Key notes that the fatal part was documented as having worn to the "max allowable limit" yet still *should have* made it to the next inspection, though it obviously didn't. He explains a crucial factor was the failure to measure the rate of wear; instead it was just pass-or-fail until the next inspection. When I asked Key if the FAA was lax in its oversight, he responded, "Absolutely."

A key player in this drama was Mary Rose Diefenderfer, an FAA veteran who in 1997 was named principal operations inspector for Alaska, until her complaints about the carrier's practices caused her to be "pushed out." She says, "That accident was a maintenance accident but in my view it was a flight operations accident, too. They were test-flying the planes with passengers on board."

Her eventual court case has filled filing cabinets. Diefenderfer's tale provides harrowing insights into not just how an airline's safety culture can be compromised, but also how the FAA itself can flagrantly fail in its oversight duties. She reported multiple violations on the part of Alaska, including a well-publicized case of the carrier's management using vodka to deice an airplane on a gate in Russia. A clever example of Yankee ingenuity? Hardly—the aircraft's auxiliary power unit was running and there were passengers on board, so only luck prevented an alcohol-fueled fire. Among her other charges were that Alaska management falsified training records, including piloting records for senior executives, and pilots were punished with time off for writing up legitimate maintenance items. In addition, low-level FAA officials were inappropriately spending time with the airline's managers, and complaints were contained within the Seattle office with the admonishment "We don't want Washington to hear about any problems." Even an FAA team later assigned to investigate this chaos confirmed that maintenance work cards were being "modified or deleted."

In return, Diefenderfer was demoted, ostracized, and humiliated. She says she was confined to a cubicle and ordered to report to a supervisor who frightened her. Within two months of her being reassigned, all outstanding items she had logged on Alaska were suddenly "cleaned up." She had been removed from overseeing the carrier by the time Flight 261 crashed; I asked her if the tragedy surprised her and she said, "I wasn't surprised by it, but I cried." She eventually received vindication in 2011, when the FAA quietly settled the lawsuit she had filed.

Such lessons were not lost on the next of kin from Flight 261, who fought for years in the courts and the press to uncover the truth. As Shaw points out, "The safety part of the FAA's mission has become very blurred. Everyone said the FAA and Alaska were in bed together." But her high-profile mission undoubtedly put additional pressure on both organizations; she recalls, "During my very angry years I said I would drop the lawsuit if [former Alaska CEO John] Kelly or one of his relatives flew on every flight. It sounded like a childish threat, but I was serious."

Meanwhile, things certainly changed at Alaska after the crash of Flight 261. A small army of feds scrutinized every aspect of the company's maintenance operation. Soon after, Alaska closed its C-check maintenance facility in Seattle and consolidated it in Oakland; then the Oakland facility was shut down. By 2005, some 92 percent of Alaska's maintenance was being outsourced. And as with all airlines, the FAA's surveillance of Alaska had shifted, with a greater emphasis on something called ATOS.

Faith in ATOS

It's called the Air Transportation Oversight System. But for years FAA inspectors have referred to the ATOS program as "A Ton Of Shit." The anger stems from a long-standing criticism: much of what is entered—or *not* entered—into ATOS is self-reported by airlines and/or is not validated. The NTSB's Haueter puts it like this: "FAA has changed its mind-set. It's all about data analysis. And I think it's garbage in and garbage out."

Even Congress worried that ATOS was becoming a replacement for the historic inspection procedure and led to an overreliance on an automated system with very little personnel input and hands-on management of the system.[4] Experts stress that data alone are not enough because data have to be validated, and that's why inspections are conducted.[5]

But ATOS has a defender in the Flight Safety Foundation's Voss: "I believe in my soul [a risk-based approach] is fundamentally correct. And ATOS is correct, too." The trick is melding electronic surveillance with traditional inspections; he says, "I think things like ATOS give you a broader picture. And your visits will produce better results."

There's a saying among FAA inspectors: Airworthiness Directives are written in blood, because ADs tend to originate from fatal accidents rather than routine inspections. If ATOS were as robust a program as was touted, then it would be generating more preemptive regulations. One congressman referred to ATOS as "a computer system to monitor a computer system." The real concern is what about the data that we now know are *not* being input?

"ATOS is a sham," says whistle-blower Gabe Bruno. "It's for their internal empire." This is not your standard employee carping; as the manager of the FSDO in Orlando, Bruno was intimately involved in a key aspect of ATOS development. In the wake of the ValuJet debacle in 1996, FAA officials claimed ATOS would be a way to provide greater surveillance of newer low-cost carriers, yet when it was conceived in 1998 only the major legacy airlines were included in the program, not the start-ups the system was designed around. Bruno began fighting for the inclusion of ValuJet (now AirTran) in ATOS, and the resistance he met from his superiors eventually destroyed his career: "That was the start of my walk into whistle-blowerdom."

Indeed, the *Seattle Times* reported in 2001 that Nick Lacey, who headed the FAA's Flight Standards Division, was removed from his post overseeing ATOS, because of "its failure to detect serious maintenance problems at Alaska Airlines before the crash of Flight 261." Bruno maintains little has changed, and points to last year's Boeing 737 debacle: "Shouldn't ATOS have captured Southwest's problems before that fuselage ripped open?"

In 2011 I asked Administrator Babbitt this, and he replied: "The most recent [Southwest] breach was not corrosion or metal fatigue, it was structural [due to its being] an aging aircraft. The airplane had been inspected and signed off. What's important about that breach is the skin ripped—and then stopped. The breach was contained and everything worked fine." On the topic of outsourcing maintenance, Babbitt said, "The good news is technology is our friend here. We probably do less monitoring on-site because of it."

Bruno's real troubles began when he uncovered systemic problems with FAA maintenance licensing in Florida, home to dozens of flight schools. He initiated retesting programs, and many "mechanics" could not pass the second exam; thirty-three of them listed their address as the same post office box in Saudi Arabia. Bruno believes exposing this scandal ended his FAA career. But the frightening coda is that individuals bearing these bogus licenses were not identified and retested for years (the FAA finally announced in 2010 that it would follow up). Meanwhile, news reports documented that one of these mechanics worked for Chalk's Ocean Airways when that carrier suffered a crash in 2005, killing twenty.

The FAA's job is all about determining risk, and that's harder with fewer fatal accident forensics. But some worry that we will go too far. Voss thinks it's all about finding the sweet spot on risk: "Are we going to become as risk-averse as Australia? They see ghosts every time Qantas pushes back late from the gate."

For Voss, the industry's long-standing policy of not publicly promoting individual carriers' safety records has harmed us all: "We've crippled aviation safety by the gentlemen's agreement, an implicit agreement not to discuss safety. So there is no incentive to improve safety." He notes how automobile makers have used safety in their marketing and advertising campaigns but warns, "Whoever is the first airline to do this will be a pariah in the industry."

Meanwhile, FAA oversight remains in question. As Brantley says, "There's an analogy with banking and finance. . . . You have an agency that really does not want to police the industry. They seem to default to the carriers." He adds, "Yes, there are risks regarding bureaucracy and complacency. But people forget what government is about: doing things we can't do for ourselves. That's the common-good part of this."

9

Threats to Survival: Why Many Air Crashes Need Not Be Fatal

APPROACH: At the end of the runway it's just wide-open field.

COCKPIT UNIDENTIFIED VOICE: Left throttle, left, left, left, left . . .

COCKPIT UNIDENTIFIED VOICE: God!

CABIN: [Sound of impact]

—*National Transportation Safety Board accident report, United Airlines Flight 232, Sioux City, Iowa, 1989*

Part I: The Most Vulnerable Passengers

Who in America is not familiar with "The Miracle on the Hudson"? When that US Airways Airbus A320 ditched in the freezing waters off Manhattan on January 15, 2009, all 155 people on board managed to evacuate without life-threatening injuries. More and more safety experts believe the happy ending was due to enhanced professionalism and advanced technology, not divine intervention. But if a miracle did occur, it was that ten-month-old Damian Sosa survived the initial impact unhurt.

"We were gliding," Tess Sosa told me. "It looks to me like we were getting down and I could see the river. And here I was with a lap child. I have a lap child and I don't know what to do. . . . In my mind it was fifty-fifty we were going to die." As we spoke, she slipped from past tense into present tense, and it was clear she was reliving those terrifying moments. Sosa told of searching for a flight attendant. Her husband and daughter were seated several rows away. And then the passenger next to her—a frequent flyer, a father of five named Jim Whittaker—turned and in "such a professional way" he calmly asked, "May I brace your son for impact?"

Sosa remembers that Whittaker placed Damian in a cradle position, with the man's knee against the seat. She heard the words "Brace for impact." Then the crash. And suddenly her baby son was being handed back to her, unhurt, and the evacuation commenced. All 155 survivors were far from safe, as ice water rapidly filled the fuselage. But both Sosa children soon were handed onto rafts, and their parents followed. And then a new journey began for Tess Sosa.

"I think it's discriminatory," she said of the FAA exclusion that requires only human beings over the age of two to be securely restrained on commercial aircraft. "How can you say a five-year-old should be restrained but a two-year-old should not?" She said she became educated the hard way, because she had no idea of the risks of holding a lap child: "It was wrong. But I'm not the only one who is accountable. The airlines need to do something about this, too."

Sosa didn't know that just one month earlier, in December 2008, a Continental Airlines Boeing 737-500 ran off the left side of Runway 34R while attempting to take off from Denver; as the NTSB noted, "the aircraft was substantially damaged and experienced a post-crash fire." There were some serious injuries but no fatalities, which was particularly lucky since there were several small children on board.

One mother, Maria Trejos, also didn't know the dangers of lap children. "It's horrible to think I could have been responsible for my child being hurt," she says. "You're not superhuman." During the mayhem of enduring the crash and exiting a burning aircraft, her husband, Gabe, held their thirteen-month-old son, Elijah, "like a football." Although Elijah had a sprained neck and bruises on his back and arms, thankfully he was not seriously hurt. And neither was Maria, who was pregnant at the time and struck in the stomach by a bag falling from an overhead bin, which almost certainly would have hit Elijah.

When I tell her of the many parents who are not aware of the dangers of flying with lap kids, she says, "They wouldn't because they haven't been through it. I wasn't thinking my child could be a projectile and slammed into the cabin. You don't think about g-forces until you're in a situation like this." None of the Trejos family has flown since the accident, opting to drive from their home in Colorado for visits to Texas.

Sometimes infants survive unscathed from turbulence, incidents, and accidents—and sometimes they do not. But widespread education about the dangers of lap children on airplanes has never taken hold, even though those dangers have been documented for decades. Sosa told me, "This is terrible to say, but maybe if Damian had perished that day something would have changed. . . . The fact that no one died really was a miracle."

The Deadly Loophole the FAA Won't Close

This loophole has an extremely long history. As far back as 1953, Civil Air Regulations section 40.174 included this language: "A seat and an individual safety belt are required for each passenger and crewmember excluding infants. . . ." Of course, fifty-nine years ago infants were not secured in cars, either, because there were no safety seats; in fact, only a minuscule number of automobiles had safety belts then. But technology and times have changed—except in the twilight zone of a commercial airplane cabin. Every state in the union has a child seat law for American roads, but there is no similar protection for American skies.

For more than twenty years, the NTSB and the FAA have been at odds—sometimes quite vocally—over this issue. In 1990, the NTSB first recommended that restraints should be required for all children under forty inches tall or forty pounds in weight; that same year United Press International reported there was "little doubt" an infant restraint proposal would soon be adopted. It's been twenty-two years, and the FAA still hasn't acted.

This all stems from what is termed the "diversion theory." As the FAA put it at an NTSB Passenger Safety Forum I attended: "Requiring the use of [restraints] for children under two would significantly increase the price of family air travel for a small targeted population." In turn, this may cause some families to drive, where the risk of injuries and deaths is statistically much higher. The FAA surmises this will lead to another sixty deaths every ten years. Therefore, due to the law of unintended consequences, the FAA has not required babies to be properly restrained.

While serving on the FAAC I had the chance to meet some of the FAA staffers who espouse the diversion theory, and I found them to be sincere and well-meaning. Their concerns over diverted highway deaths cannot be dismissed out of hand. But fundamentally I disagree with the logic. If buying a ticket for a small child will force some families to drive and therefore endanger their lives, then by extension *any* fare increase would have the same effect, so why limit the argument to two-year-olds? Adults are in danger as well, yet I certainly don't envision a post-deregulation DOT that would dare to ban airline price hikes for fear some passengers will drive.

What's more, the FAA's reasoning is based on old and suspiciously faulty data. The diversion argument was fashioned in the 1990s, before low-fare carriers remade the industry. Questions abound: At what price threshold would some families divert to driving? What about long-haul and over-water routes where driving is not an option? And how certain are we that parents will divert in the first place? As Tess Sosa puts it, "What parent in their right mind would say, 'I'm not going to pay three hundred and fifty dollars?' You're taking a risk. My belief is this diversion theory is just a smoke screen."

She is not alone. NTSB chairman Hersman says, "I think the diversion argument is completely spurious. It's a statistical stalemate." She clearly feels the discussion has ended: "In an industry as safe as aviation, and one that puts such a premium on safety, it's an embarrassment. It's past time. . . . No one who works in safety will tell you that restraint use is not important. . . . I have to really question why we take such an approach with the most vulnerable passengers."

Once Maria Trejos recovered from the Denver accident, she found fault with the FAA's reasoning as well: "To me, the cost of my child's safety outweighs the cost of the [airline] seat. . . . If the auto industry can [mandate] it, why can't the aviation industry?" She had checked Elijah's child seat as baggage, and when she later learned about the FAA's diversion theory, she felt it was all about money, particularly since many road trips can cost just as much for gas and lodging. Trejos notes that since the airlines charge extra for so many other things, they could find ways to sell seats to infants.

In this age of ancillary revenue, one would expect the airlines to embrace the chance to sell additional seats for babies. In fact, most major carriers do offer discounted tickets for infants on international flights, and Southwest deserves credit for offering such seats on domestic flights. The irony is that in this era of fee-happy airline executives, carriers that decide to rent safety restraints stand to make even more money, but the FAA refuses to change its position.

A Modest Proposal: Ban Lap Kids

Shortly after I was appointed to the FAAC's Safety Subcommittee, I formally recommended that the FAA mandate the use of approved child restraint systems (CRSs) for all passengers under two. I also officially requested the statistics on lap children injuries and deaths on commercial flights. The answer: over the last twenty-five years there have been three infant fatalities that otherwise would have been survivable (those terms *survivable* and *nonsurvivable* can be quite subjective, though here it's clear infants died due to a lack of CRSs). But the shocker was that the FAA was unable to tell us how many lap children have incurred injuries. I was dumbfounded the FAA did not possess these numbers, but we were assured further research would be conducted.

"I'd like to see statistics on how many babies have not become functioning adults because of injuries," says FAA whistle-blower Kim Farrington. She notes that one major carrier paid millions after an infant flew out of a guardian's arms and hit an overhead bin. She adds, "The FAA is catering to the desires of airline executives."

After the FAAC asked me to convene a meeting of the nation's experts to further research the topic, we all gathered outside Washington in October 2010: representatives from the FAA, NTSB, DOT, Air Transport Association, Boeing, the Association of Flight Attendants, and the airlines. I opened by asking if *any* experts believe a lap child is as safe as, or safer than, a child in a CRS. And the answer was a unanimous no. This is an issue without pending studies, suspect science, or warring camps. Science has proven that lap children are at greater risk of injury and death, not only during accidents but even during routine turbulence. This discussion begins and ends with the physics, and the physics is irrefutable. As the FAA itself states on its site: "Did you know the safest place for your little one during turbulence or an emergency is in an approved child restraint system or device, not on your lap?" So the "debate" is not over safety but over economics and unintended consequences. Yet there are other factors: the whisper campaign that focuses on FAA indifference and the airlines' concern that charging for infant seats will turn away customers.

More so than any other aviation safety topic I research, more so than outsourced maintenance shops in El Salvador or sleeping air traffic controllers, the issue of unrestrained infants invokes the mantra of the FAA being the "Tombstone Agency." A longtime airline official says simply: "Not enough babies have died. They [the FAA] deserve being called a tombstone agency."

One of Hersman's predecessors as chairman of the NTSB, James Hall, believes there's a political component: "It's clear that in government very little attention is paid to those with no voice—the elderly, those with low incomes, and children."

To its credit, the FAA has attempted to heighten awareness; on Thanksgiving weekend 2010, the agency's home page carried warnings about lap kids, and the FAA has begun utilizing social media and parenting sites. Former administrator Babbitt told me, "We've done a lot of research. We're reminding people. We're seeing a lot of compliance. . . . And we're continuing to collect data." He added that these data are "mixed." Meanwhile, I flew on dozens of flights to research this book, and in some cabins I counted three, four, or even more lap kids.

The Arguments Against Securing Infants

I've written about this issue extensively, and each time I do I'm subjected to a higher than usual amount of nasty (and uninformed) blog comments. It strikes me how often readers respond in the vein of "If it's so dangerous, why do the airlines and the FAA allow it?" Why indeed.

But there's no shame in being unaware of the dangers. Even NTSB chairman Hersman advises, "I was one of those parents who flew with a child on my lap. Then I came to the Safety Board and I learned. . . . People look to the government for laws and guidance." Hersman points out that compliance with state child restraint laws in automobiles exceeds 95 percent, even where adult compliance lags.

After my last column for USAToday.com I wrote a follow-up responding to the standard arguments against mandating child restraints. In addition to diversion, there are four other claims that surface time and again.

The "Superman" Argument

It's all about the g-forces, and no one this side of the planet Krypton can argue with physics. The most loving parent or caregiver in the world can't hold on to even the lightest infant in an aircraft traveling at four-fifths the speed of sound. What's more, the weight of the adult can crush a baby held either inside or outside the adult's seat belt. This is the argument that isn't an argument; anyone who claims a child is safer in their arms is dead wrong.

The "We're All Gonna Die Anyway" Argument

Why worry about child restraint systems? Or air bags? Or even seat belts? If anything goes wrong on an airline flight, it's all or nothing—and no one will walk away. This old and tired argument has always been wrong, but in recent years it's become *very* wrong. Also, CRSs prevent injuries and deaths under conditions that pose much less serious threats to adults, such as severe turbulence.

The "Nanny State" Argument

This is a corollary to the "get government off our backs" contention. Legally, politically, morally—this argument is bogus. Requiring that children under two be strapped into seats on a commercial aircraft traveling at .82 Mach is not the intrusion of a nanny state bureaucracy dictating how kids should be raised or fed or disciplined. It's the closing of a loophole that has existed since 1953. In fact, by this reasoning *all* passengers should be free to remain unbuckled or even unseated on commercial airline flights. The FAA has a congressionally mandated responsibility to ensure the safety of all of us. Furthermore, to reduce the risks to their crudest elements, a flying projectile—even a human one—poses a hazard to everyone else in the cabin, so closing the loophole increases the margin of safety for everyone on board.

The Statistics Argument

What we're talking about is a handful of deaths and an unknown number of injuries. This is what is whispered in Washington, but no government official or airline executive will dare speak it aloud. And yet it's the most persuasive argument for doing nothing and allowing FAA and airline inaction and a gross lack of education to endanger more and more infants every day. But it's best to raise this claim—how else to refute it? As passenger advocate Kevin Mitchell puts it, "To be cornered on this as a statistical analysis is to lose the argument. We're a country that sends out helicopters to save a dog in a swollen river. No one questioned the cost of saving the Balloon Boy in that moment. Do we really want to look at this as only a dollars-and-cents issue?" Ultimately, that's the only question that matters.

Falling Through the Safety Net

Unfortunately, not all lap children have been as lucky as the Sosa and Trejos babies. And no one knows this better than Jan Brown, a flight attendant on board United Flight 232 when it crash-landed in Sioux City, Iowa, in 1989. Over the years she has told and retold the harrowing tale of how she instructed the passengers to brace for impact, and as per standard operating procedures, told a mother to place her lap infant on the floor. As the NTSB accident report noted: "The mothers of the infants in seats 11F and 22E were unable to hold onto their infants and were unable to find them after the airplane impacted the ground." Minutes later, Brown had the heartbreaking duty of confronting one of those mothers outside the burning wreckage and informing her that her child had died.

That experience transformed Brown into a crusader for banning lap kids, particularly after the NTSB report made that recommendation official. At an NTSB briefing she testified: "When preparing the cabin for an emergency, flight attendants should not have to look a parent in the eye and instruct them to continue to hold the lap child when we know there is a very real possibility that child may not survive without proper restraints. . . . No parent should find out in this way that holding a child on a lap is unsafe." I asked Brown if she ever imagined this issue would be debated twenty-two years after the crash. "I'm very surprised," she answered. "God has a funny sense of humor to have someone who is afraid of public speaking talk about this for so many years."

Her passion came through as she recounted the long battle she has fought: "They're looking at it one way. I'm looking at it as a flight attendant who had to face a mother whose child had died. They want us to tell everyone over two to brace for a crash. But those of you under two—well, tough. Thanks for flying with us . . . The FAA operates on a body count. . . . That's the FAA mentality. It's criminal. Then why do we need an FAA?" Brown said she has found peace by determining such matters are in God's hands. But she thinks of the child killed in Sioux City and says, "I can't let that little boy alone. I'm tenacious. I finish what I start."

In September 2011 I traveled to Washington to receive an update from the DOT on the FAAC's proposals. I was tremendously disappointed to learn that the FAA has no immediate plans to mandate child restraints and instead will continue to educate parents about the dangers of lap children, a formidable challenge indeed. I asked an FAA official about this, and he responded in two words: "Cost-benefit."

Part II: It's No Longer All or Nothing at All

Experts point to multiple serious commercial accidents in recent years in which there were no fatalities. This list keeps growing, and includes four rather extraordinary recent cases:

- August 2005: An Air France Airbus A340 with 309 people on board ran off the runway upon landing in Toronto and burst into flames; there were no fatalities for the flight dubbed the "Miracle in Toronto."

- December 2008: The Continental Boeing 737 carrying 115 people ran off the runway upon takeoff in Denver with the Trejos family on board; although the plane was severely damaged and there was a postcrash fire, thirty-eight passengers and crew were transported to hospitals—but no deaths.
- January 2009: The US Airways Airbus A320 with 155 people on board—including the Sosas—ditched in the Hudson River in New York City after a debilitating bird strike disabled both engines; there were no deaths on the "Miracle on the Hudson" flight.
- December 2009: An American Airlines Boeing 737 carrying 154 people overran the runway while landing in Kingston, Jamaica; although the aircraft fractured, there were injuries but no fatalities.

In all four cases, the aircraft were destroyed; two were burned beyond repair and one was dragged from a river. Yet a combined total of 733 people lived to talk about their experiences—and not one succumbed. "They're not miracles anymore," says NTSB chairman Hersman. "We see these miracles more frequently than the crowd killers. It's a myth that is perpetrated." On this issue she is in agreement with former FAA administrator Randy Babbitt: "The Miracle on the Hudson, that was no miracle. It was a symphony."

It's past time to jettison the tired mantra that no one walks away from a plane crash.

Damn Statistics and Lies

Surviving even the most grisly of crashes is not theoretical. One need only look at United Flight 232, which experienced a "catastrophic failure" of the tail-mounted number-two engine, causing high-speed shrapnel to sever all three hydraulic systems powering the DC-10's flight controls. Through a remarkable display of piloting skills, Captain Al Haynes and crew utilized only engine power to "steer" the plane to a rough and fast descent into Sioux City. The subsequent crash killed 111 of the 296 people on board, though many airline veterans would have said a total loss of flight controls was a nonsurvivable scenario.

A native Texan, Al Haynes naturally sounds like an airline pilot cast by Hollywood. He learned to fly as a Marine Corps aviator, and subsequently accumulated more than twenty-seven thousand hours of flight time on five different aircraft during a thirty-five-year career for United. He's almost eighty now, but he remains sharp—and humble. "I was just in the wrong place at the wrong time," he says modestly, though aviation experts know better. For more than two decades now, aeronautical types have spoken in awe of the tremendous piloting performance of Haynes and his crew, long before Captain "Sully" Sullenberger justifiably became a coast-to-coast hero for his own tremendous feats on the Hudson River.

Haynes has earned the right to decry the "all-or-nothing" mantra by calling it "a pessimistic attitude." But he places partial blame on media coverage: "The media push it when there are injuries and deaths, but not when there aren't. People very quickly forget there are survivable accidents." He praises Sully for the "fantastic thing he did," but adds that had there been fatalities on US Airways Flight 1549, "it would have been an even bigger story."

Such criticism of media coverage may sound like sour grapes; after all, what editor or producer would *not* lead with a plane crash that killed one hundred or more people? (NTSB members often say, "If it bleeds, it leads and if it scares, it airs.") But more than two decades ago, an expert studied the front page of the

New York Times for two years and found the following: "Page-one coverage of airplane accidents was 60 times greater than reporting on HIV/AIDs; 1,500 times greater than auto hazards; and 6,000 times greater than cancer, the second leading killer in America after heart disease."

That expert is Professor Arnold Barnett. It may seem odd, but one of the nation's leading authorities on aviation safety statistics struggled with aerophobia himself. In his office at MIT, we discussed the rumor. "Yes!" he said. "It may not be a total coincidence." He pointed out that when he was growing up in the 1950s, airplane crashes were rather common, and by the time he flew for the first time in 1967, he was more than nervous. He wondered how he could sublimate these fears, and struck on the idea of analyzing airline safety records, noting, "There really had not been a formal statistical analysis."

Barnett asserted that the death risk in boarding a commercial U.S. airline flight now is about 1 in 20 million, and stated he is "in awe" of the airline industry's safety record: "It's sort of like the Eighth Wonder of the World."

Fearful Flyers and Fearful Airlines

I've never suffered from aerophobia (I never would have lasted at Overseas National Airways or Tower Air if I had). A few years ago, shortly after I started teaching at Hofstra University, I stumbled upon a leading authority in aerophobia and saw his email address ended in "@hofstra.edu."

Mitchell Schare teaches psychology and founded the Fear of Flying Treatment Program at Hofstra's Phobia and Trauma Clinic. I'm a Tuesday/Thursday guy and luckily he is, too, so I trekked cross-campus and strapped myself in for a multisensory virtual reality ride. I assumed it would be rather cheesy but found it surprisingly realistic, even though my blood pressure didn't rise at zero hour. The most important thing I learned is that the generic term *fear of flying*—aerophobia—actually is an umbrella term for at least seven separate fears, with subsets of fears to boot. Passenger A may be afraid of heights and Passenger B may be unnerved by loss of control. What's more, one person may be bothered by claustrophobic germy air while another is freaked out by claustrophobic physical closeness to others; the potential combinations are extensive. "Naïve people assume everyone with aerophobia is afraid of dying," explained Schare. "But not everyone is afraid of dying."

Because so many fearful flyers need to medicate and/or intoxicate themselves in order to board an airplane, that poses additional risks if that airplane needs to be evacuated—when every additional second literally threatens the lives of others. "In terms of survivability, you need the crew and the passengers to work together," Schare said. "People have to follow instructions and someone has to take control. Who will survive? Some people will be on the program and some will lose it." He explained that owing to dissociation—when certain individuals lose direction during a life-threatening situation—some passengers may act as if in slow motion.

Of course, the FAA requires the preflight safety briefings, but the redundancy prompts some of us to tune out. That's why Schare gives credit to Southwest for allowing flight attendants to joke during the briefing; a common bit is, "If you're securing the oxygen masks of two children, decide which one you love more." He explained that this forces even the most frequent of flyers to suddenly pay attention during a canned speech they may have shut out. Similarly, Virgin America has created a clever in-flight video.

Experts say the airlines don't like publicly addressing this topic, over concerns that skittish passengers will opt not to fly. Schare, who forces his fearful patients to confront their worst terrors, applies the same principle to the aviation business by noting, "You're talking about the big scary thing the airlines don't want to talk about—a crash." He maintained that more can be done: "I think the airlines need to step up and address this."

Former NTSB chairman James Hall says fault lies with the industry: "The airlines will not do the basic education that is needed. Because they're concerned about the adverse economic impact on the bottom line." But he asserts the ultimate responsibility lies with the FAA, and says, "I believe in the American system of government, even though it's not always implemented. Many times safety has to be the responsibility of the government because some corporations view death as just an economic factor rather than as a tragedy."

Learning from Tragedy: Technological Improvements

"I felt the heat from the fire and I thought, this is the end," Jan Brown recalls of the Sioux City crash. "It was so peaceful and painless." Then reality struck that she was alive, and her instincts kicked in: "Thank goodness for that training." She is echoed by her former captain, Al Haynes: "It was all due to good training." Brown recalls she wasn't sure if the aircraft was losing pressure and she would be sucked out of the fuselage so she sat down in the aisle and held on to a seat support; although she lost consciousness, she didn't loosen her grip, knowing she had only sixty to ninety seconds to get out.

Nora Marshall, chief of the NTSB's Survival Factors Division, maintains that emergency training for crew members—coupled with good communication—is all-important: "There's really been a lot of evolution in protecting passengers." That's why Brown has real respect for the performance of the flight attendants on the Hudson River, who executed a near-flawless evacuation: "It was incredible that people got off with their life vests on. I was in awe of that. I know how tough it is to get them on."

Preventing such occurrences in the first place is even tougher. Kendall Krieg, senior manager of flammability certification for the nation's largest aircraft manufacturer, says, "The number one thing we focus on at Boeing is eliminating the event to start with." On my visit to the company's massive facility in Renton, Washington, he and his colleague Al Carlo, a safety airworthiness manager, spoke at length about accident prevention and avoidance. The list of technological and human factor advancements in commercial aviation is impressive, and could easily fill another chapter of this book. But I asked them to address the worst-case scenarios, and accident survivability.

"We've developed more fire resistance," said Krieg. "Trains and cars don't come close. We work to a much higher set of standards." Those standards stem from the industry's ninety-second rule, which requires that all occupants be off the plane within the "golden time" of a minute and a half. Carlo explained that there are fire detection and suppression systems in the engines and cargo areas, but in the cabin the focus is on postcrash. This means using better materials in seats, carpeting, bulkheads, and bins, so the hazards of both flammability and smoke inhalation are held off long enough for all to get out. He pointed to the 737 in Denver; in that case the metal itself burned before the interiors, allowing all to escape that aircraft.

Corky Townsend, Boeing's director of aviation safety, explains that her fellow engineers work toward three main goals for occupants: surviving an accident's impact; surviving a postcrash fire; and safely evacuating. She also states there have been marked improvements in evacuation slides; the historical failure rate has been so high that during drills the industry assumes that 50 percent of all doors and/or slides will not be available. An important point: the NTSB estimates that evacuations occur on U.S. aircraft once every eleven days on average.

But improving aviation safety through advanced technology is a never-ending pursuit, and one way to enhance survivability rates is to better understand fatal accidents. For example, the NTSB's "Most Wanted" list of improvements in aviation includes installing "crash-protected image recorders in cockpits to give investigators more information." What's more, Star Navigation Systems of Canada has developed an in-flight safety monitoring system that uses Global Positioning System tracking software to record the information currently stored in flight data recorders and cockpit voice recorders (the orange boxes the media refer to as "black boxes"). Considering that the recorders from Air France Flight 447 were found in the spring of 2011—nearly two years after the Airbus A330 disappeared over the Atlantic Ocean—the concept of transferring such vital data to ground-based systems could greatly aid accident investigators.

Simple Steps That Could Save Lives

Of course, some types of airline accidents remain nonsurvivable, and it's likely that will never change. In fact, Barnett's research has unveiled a key point: approximately 80 percent of passenger fatalities occur on flights where there are no survivors. But the good news is quite good: experts now say that 90 percent of all commercial aviation accidents are survivable. In fact, an exhaustive NTSB study that focused on all domestic airline accidents over seventeen years found that overall, 95.7 percent of occupants survived.

Having undergone emergency egress training at Tower Air and again in the Air Force Auxiliary, what I learned affects me on every flight I take. That's why experts say that passengers themselves have a big part in determining who will live and who will not. Captain Haynes says: "I tell people all the time: It's your own personal responsibility to get out of the airplane. You need to think about where the exits are and what you'll do."

Wearing the Right Clothes

The TSA has invoked and revoked numerous policies since its formation in 2002. It has even flip-flopped on flip-flops; a few years ago the TSA suggested: "Footwear that screeners are less likely to suggest you remove includes 'beach' flip-flops or thin-soled sandals." That advice was in direct opposition to the FAA's recommendation to wear shoes that won't easily slip off on an evacuation slide (the FAA and TSA have since synchronized their message). However, shoes should not be high heels—especially spiked heels—because they will likely cause a broken ankle on that slide.

The flip-flop issue underscores an important point: too many passengers dress for the destination rather than the departure. The flight may be headed to Nassau or Maui, but the departure may be aborted in snow, ice, rain, or mud. Those 155 people in the freezing waters of the Hudson River can affirm that. Jan Brown invokes the terror of Flight 232 and recalls, "I saw that one man's shirt was burned right off him and I remember thinking: polyester." By contrast, another passenger had dressed safer by choosing all-natural fibers such as wool slacks and cotton socks. Brown herself suffered second- and third-degree burns on her ankles from her flammable pantyhose, and stopped wearing skirts and dresses on airplanes after that.

It's about common sense, not investing in a new wardrobe. Keep shoes on during takeoff and landing, and make sure they are flat-heeled shoes that won't fall off easily. Avoid highly flammable fibers, and think about the outside weather; a tank top in New England in February doesn't make sense.

Choosing the Right Seat

It's not preordained that all airplanes will respond in the same way to a given accident scenario. The industry is filled with pilots and mechanics who swear by one particular manufacturer or model. Crew members can be spotted carrying airline-issued cases known as "brain bags" through airports with the bumper sticker: IF IT AIN'T BOEING, I AIN'T GOING.

Then there's the size issue, which like so much else about aviation safety ultimately comes down to a matter of physics. "Basic awareness will help you out," says Todd Curtis. He points out that the larger the aircraft, the greater the chance of survival, and suggests if the Airbus A340 in Toronto had been a DC-9, many might not have lived. On the other hand, larger passenger totals raise the evacuation X-factor, as was seen in 2005 as Airbus executives sweated the safety trials of the A380 superjumbo, when 873 people were given ninety seconds to evacuate.

While some factors are beyond the passengers' control, choosing a seat is not. It's important to select a seat in an emergency exit row, or close to an exit or multiple exits. As aviation expert Mary Schiavo notes, the row behind is better than the row in front, primarily because in an emergency people will instinctively move forward.

For decades myths have sprung up in airline circles that passengers in the rear of planes have a better chance of surviving. This stems in part from the obvious assertion that an aircraft moves at great speeds in a forward direction (hence the old axiom "You can't back into a mountain"). Also, in many cases the rear empennage sections of airplanes—which on most commercial planes today don't house engines and fuel tanks—often break away from the fuselage. So I put it to the experts: Are some rows or seats statistically safer than others? "I don't believe one seat is safer than the others," said Dunham. MIT's Barnett agreed: "One can get too obsessive about it. In some ways it really is all or nothing." Boeing said it has not explored the issue.

As for rear-facing seats, the NTSB's Nora Marshall notes the issue of debris flying forward in an accident scenario. She also explains that such configurations require "much, much stronger floors," which of course add weight. So like so much else about the airline industry, there is a cost component. In 2005, the FAA implemented the "16g Rule," requiring that passengers be able to withstand forces equal to sixteen times earth's gravity (unrestrained infants, of course, were exempted). Engineers soon realized that meeting this new requirement would require removing seats from cabins to expand the distance between rows or installing "three-point" shoulder harnesses, which require reinforcing cabin floors. The airlines hated both ideas because removing seats and adding weight are costly solutions. So a third alternative was developed by Phoenix-based AmSafe: an aircraft airbag built right into the nonbuckle end of a passenger's seat belt. "Over five to ten years, it will be as ubiquitous as seat belts are today," says Bill Hagan, president of AmSafe. But Corky Townsend of Boeing states, "There are some concerns with failure rates."

Buckling Up

Overwhelmingly, most airline accidents occur either during the initial takeoff stage or the final landing stage of flight; there's a reason that seat belt sign is lit at those times. A Boeing study of accidents and fatalities examined commercial jet aircraft crashes between 1959 and 2010, and adjusted them for an average flight time of ninety minutes; the findings for fatal accidents:

- 17 percent occur during takeoff and initial climb (2 percent of total flight time)

- 5 percent occur during flaps-up climb (14 percent of total flight time)
- 16 percent occur during descent and initial approach (23 percent of total flight time)
- 36 percent occur during final approach and landing (4 percent of total flight time)

In fact, just 11 percent of fatal accidents occur during cruise, which represents 57 percent of total flight time for a ninety-minute flight, and obviously a much longer period for lengthier flights. The good news is that survivability increases when airplanes are closer to the ground, because they're operating at relatively lower speeds and often there is more maneuverability to land the aircraft safely near an airport. And airports, of course, have trained crash and rescue crews standing by for just such events.

That said, buckling up may be the simplest of all smart steps to take. But a veteran pilot believes crew members can do more to educate passengers: "The crew should explain when the sign is on. The passengers don't understand why. Often there is poor communication between cockpit and cabin crew, and between crew and passengers." One expert cites United Flight 811 in Honolulu in 1989 as a striking example; when a cargo door blew out at twenty-three thousand feet, the "explosive decompression" sucked out nine people, including one passenger not buckled in.

Securing All Loose Items

Schiavo offers explicit advice for passengers: "Do not let anyone block your exit by overstuffing the underseat area so it blocks the row." She also advises monitoring what goes into the overhead bins, to ensure the items are not too heavy and not blocking good closure of the latch. That image of a pregnant Maria Trejos being struck in the head and stomach by baggage when Flight 1404 crashed in Denver serves as a reminder. She feels the airlines and FAA could do more to educate passengers, and recalls a fellow passenger evacuating the plane on a cold night in December without her glasses or shoes because she had taken them off prior to the takeoff roll.

Avoiding Alcohol, Drugs, and Sleeping at Key Times

Dianne McMullin, an associate technical fellow at Boeing and an expert in human factors and aviation safety, examines airplane accidents from a physiological and psychological perspective. She also notes an industry that by definition is global in scope will encounter cultural and language challenges that make rapid evacuations much tougher.

Then there is the drinking issue. Several studies have been conducted to determine if the effects of alcohol are exacerbated at high altitudes, and while the results have been mixed, medical experts do agree that alcohol acts as a diuretic and will exacerbate the dehydration effects felt when airborne. Simply put, there is no denying that drinking and flying don't mix well. At a minimum, McMullin notes that alcohol impairs judgment, slows reaction time, and induces drowsiness: "The combination of all these things and putting on an oxygen mask is not a good thing."

However, this stark reality conflicts with the airline industry's newest sacred cow: ancillary revenue. Most carriers are in the bartending business, so with all that cash flowing in, they're not about to declare last call. It's worth noting that flight attendants on board Virgin Atlantic Airways include a warning about the dangers of consuming alcohol at high altitudes; then again, the carrier's site notes there is "a free bar service" in economy class. Coincidence?

Monitoring Situational Awareness

Boeing's McMullin observes that most human beings "overestimate the probability of good things and underestimate the risks."

While flying needn't be a morose undertaking, staying aware of surroundings is all about establishing good habits on board an airplane. It means paying attention to all instructions and advisories from crew members, and shutting off headsets and iPods so you can listen during takeoff, landing, and other critical phases of flight. It also means counting the number of rows to the nearest exits, in both directions, and not removing shoes and eyeglasses before takeoff. In addition, some passengers may want to invest in a smoke hood, a proposition Ralph Nader has supported for years.

Captain Haynes notes the dangers of frequent flyers who mistakenly believe they know it all: "We've become so blasé. Even if you fly a lot, every airplane is different."

Preparing for the Unexpected

In addition to a lack of passenger awareness, two trends are working against faster egress from commercial airplanes. One is that larger Americans are being squeezed into smaller airline seats. A 2001 study by a British ergonomics firm found that most economy-class seating dimensions provided hindrances to egress, and that was *before* planes became fuller, seats became even smaller, and Americans became even bigger. The second factor is those ever-increasing passenger load factors; an obvious corollary is the fuller the flight, the greater the chance not everyone will safely escape the airplane. Haynes knows it firsthand: "The high loads make it tougher to get off."

10

Lights, Camera, Strip Search: The Tragicomedy of Airline Security

> You should neither be asked to nor agree to lift, remove, or raise any article of clothing to reveal your breast prosthesis, and you should not be asked to remove it.
>
> —*TSA.gov*

AND THEN THERE WAS THE time I was convinced I was the only United States citizen in the world who knew the details of the next terrorist attack against America.

In early 1989, Tower Air sent me on an extended worldwide road trip to oversee a "wet-lease" operation for London-based Air Europe, a carrier growing so fast that it literally did not have enough planes and pilots to operate all its flights. My job was to oversee the airport handling. I spent some time living in a bed-and-breakfast in Crawley, just outside London's Gatwick Airport, while ironing out details at Air Europe's headquarters. Then it was on to the Dominican Republic, Barbados, and Bangkok, before settling into Bahrain.

A tiny island country in the Persian Gulf, Bahrain is about three times the size of Martha's Vineyard if you're measuring in square miles, but during the oil boom it was one of the richest per capita nations on earth. It's connected to Saudi Arabia via the King Fahd Causeway and we soon learned that many devout Saudis often crossed that bridge on Friday evenings, because the hotels in Bahrain offered everything from alcohol to gambling to prostitution—and apparently what happens in Bahrain, stays in Bahrain. Later, during the First George Bush Gulf War, Bahrain warmed up to America and the island became a de facto U.S. air base. But in 1989 relations with Uncle Sam were still pretty chilly, as I was soon to find out.

It's worth noting that I became the manager of worldwide ground operations for Tower Air when I was just twenty-six. On most of my international trips I traveled without business cards or even a credit card, and sometimes without a pen. My only supervisory experience was as the assistant manager of a gas station on Long Island. But there I was, the face of Tower Air—and sometimes even America—around the globe. That year I routinely provided my signature for more than $1 million in billings for jet fuel alone.

On my second day in country, I met with Bahrain's aviation ministers, carrying a lengthy legal pad list of THINGS TO DO TODAY. I entered a plush room on the roof of the airport and found myself at a table with ten other men. I also noted the obvious, that I was the only one not wearing a keffiyeh on my head. But the meeting went well and I was treated cordially. We secured arrangements for fueling, baggage handling, and light maintenance of the Tower Air 747 that would be carrying five hundred Air Europe passengers from London.

Then things took an odd turn. Suddenly the head man, who was effectively Bahrain's secretary of transportation, smiled and asked if I was American. I told him I was. Interesting, he responded, an American working for a British airline? I explained that I worked for the airline actually operating the flights for the next several weeks. Twenty eyes turned toward me. And what airline is that? "Tower Air," I replied.

Ten mouths opened and ten barely audible gasps could be heard in unison. The aviation minister stared at me. "That's a Jewish airline," he said. "No," I replied, thinking that perhaps I could go for a quick laugh and change the subject. "In the United States airlines don't have religions." The room was not amused. And the aviation minister continued to stare. "You know what I mean," he told me.

I did know. He was referring to Tower's close ties with Israel, particularly since our chairman and senior executives were Israelis, and former employees of El Al. In fact, virtually our entire security staff had come over from Israel's flag carrier. The JFK–Tel Aviv route remained Tower's bread-and-butter source of income. In response, I shifted gears. There was plenty I didn't know about finance and management, but I thought I could reason my way out of what was quickly developing into a bit of a sticky wicket, as the folks back in Crawley might say. I quietly explained that Tower Air was a U.S. flag carrier, headquartered at JFK in New York City and incorporated in Delaware. We flew the Stars and Stripes on the tails of our planes (though in some countries we whited out the flag). And we transported U.S. military troops around the world.

The aviation minister shook his head. "You work for a Jewish airline," he said. Thankfully I averted disaster by not laughing, because I instantly and regrettably recalled the scene in the movie *Airplane!* in which an El Al aircraft taxis in with pais ringlets attached to either side of the cockpit. I took a deep breath, unsure of how to proceed, and quietly pointed out that a Tower 747 was scheduled to arrive in Bahrain in about seventy-two hours. One of the other men at the table, who looked as if he could have been a relative of the aviation minister, suddenly spoke up. "If that airplane attempts to land here this week," he told me, "it will be blown out of the sky."

Talk about a deal breaker. I attempted the oldest stall tactic in the book, by politely asking the gentleman to repeat himself. But he would have none of it. There was no question that I had heard him. For a few long moments, I sat in silence, my stomach churning and my ears pounding. I felt as if I had intercepted a telegram on December 6, 1941, and the news out of Tokyo wasn't good. Was I really the emissary being warned of the next major terrorist attack on the United States? It was April 1989, and the recent history of the industry I was working in wasn't pretty. U.S. aircraft and cruise ships had been targeted for hijackings throughout the 1980s, and at that very moment they were still collecting pieces of the Boeing 747 that had operated as Pan Am 103 and been blown to bits over Lockerbie, Scotland, that December. The implications were staggering. There was a new president in the United States, and new presidents were known to assert themselves early in their terms. Would the world be witnessing a cataclysmic war by week's end? And was Bill McGee really the only American who knew what was to come?

My head hurt. Severely. But as I stared at all those unhappy faces, I was suddenly struck by a simple idea. Why not table this item—you know, the unsettled matter of the 747 with more than five hundred souls on board being blown to proverbial smithereens. Instead, why not move on to the other bullet points on my legal pad. For instance, who exactly would be dumping and replenishing the aircraft's lavatory fluid? And what about fresh ice for the crew meals?

Somehow I made it through that meeting, gastric juices stirring and cranium pounding, and within an hour or so I was free to leave. I shook all ten hands and then raced downstairs and into the oppressive Persian Gulf heat. I was already sweating when I caught a taxi back to my hotel. Once there, I raced upstairs and calculated the time difference and placed a call to the States. Not to Tower Air's offices, but directly to the chairman's home. Morris Nachtomi answered from a sound sleep, and I identified myself.

He owned a small and troublesome global airline, so he was used to such calls. In fact, I had woken him more than a dozen times myself. "Yes, Bill? What is it?" I felt relieved, knowing that soon I would not be the only person from the Western Hemisphere carrying this terrible burden. Was my phone being tapped? Fuck them, let them tap it. Anything to prevent our aircraft, tail number N602FF, from making history. "We've got a situation here. A very serious situation. We're going to have to cancel all the Air Europe flights."

Nachtomi was an easy guy to anger but a hard guy to rattle. "Why is that?" I paused. "Well, I just came from a meeting. With the aviation authorities. The ministers. And . . . and, well, they told me that when six-oh-two comes in this week . . . to Bahrain . . . well, they're going to shoot it out of the sky." I instantly felt better for having shared the grim news.

Nachtomi sighed, and that sigh carried quite audibly all the way from Westchester County. "Is that it?" he asked. *Is that it??* "Well . . . yes," I answered. "Bill," he explained patiently, "they *always* say that in Bahrain."

Oh. Well, who knew? The proverbial rookie mistake! Finally I blurted out, "They seemed pretty serious." But Nachtomi had already moved on. "Is there anything else, Bill?" Anything else? You mean, anything besides the pending international terrorist incident, the imminent explosion of a U.S. commercial jet, the killing of more than five hundred innocent civilians? "No, not really," I said quietly. Thus my introduction to the world of airline security.

Welcome to the Show

For years a cult of sci-fi readers has worshipped Philip K. Dick's 1974 police state novel *Flow My Tears, the Policeman Said,* in which paramilitary law enforcement officers in Los Angeles arrest citizens without valid ID at impromptu security checkpoints. Dystopian fantasy? Maybe in 1974. But not so much for Americans venturing outdoors after 9/11, who regularly encounter such ad hoc screening in train stations, office buildings, and baseball stadiums. (In 2006—in an aviation irony way too extraordinary for fiction—Comic-Con organizers ordered a remote-controlled android of Dick, but en route to San Diego the folks at America West Airlines misplaced the android's head.)

Of course, androids aren't the only ones losing their heads over aviation security. It's a murky and at times bizarre world, and many of the people I speak to are characters that would fit in at Comic-Con. When I tap into the security network, my inbox starts filling with anonymous emails. I'm drawn into internal feuds. One expert asks me to pay him for his opinions (I decline). Another urges me to carry a pineapple taped up with wires in my checked suitcase to demonstrate how poorly bags are screened. Some question my patriotism for even writing about the Transportation Security Administration, while others question my patriotism for not writing about it more often. But one belief unites nearly all the professionals: the TSA—which they maintain stands for "Travelers Standing Around"—is all about security theater, not security. And passengers are just unpaid extras in this drama.

For many of us, this show has gotten old. As F. Scott Fitzgerald wrote: "So we beat on, boats against the current." We remove our jackets, our shoes, our belts, our dignity. We glance at a June 2011 headline from CNN.com: "TSA Denies Having Required a 95-Year-Old Woman to Remove Diaper." We raise our hands while complete strangers with no medical training prod our most sensitive regions. And we note that confidence in the system was not rewarded in December 2009, when news broke that the TSA had accidentally leaked its airports screening manual via the Internet.

What it all comes down to is risk. And just as aviation safety experts debate the acceptance of risk, so too is aviation security inherently risk-based as well. As travel columnist Joe Brancatelli says, "With both safety and security, we've created a zero-based model. That's the TSA and it's a nightmare."

Of course, any discussion of risk becomes a discussion of cost, and there's no doubt the airlines are still concerned about the cost of security—whether it's paid by the carriers themselves or paid by passengers via that $2.50 "enplanement" September 11 security fee. Sometimes the squabbling between the industry and the TSA spills over into public, as in October 2010 when the Air Transport Association openly asked that air marshals not occupy first-class seats, acknowledging that the airlines' concerns were triggered by the loss of revenue in profitable first-class sections.[1]

As a nation, we haven't had a robust discussion about security or risk, let alone how they affect aviation. Instead classified information has been doled out through the highly questionable color-coded Homeland Security Advisory System and its successor, the National Terrorism Advisory System. And all the important decisions are made for us.

"Virtually everything we do is based on illusion and there is very little substance," says Bogdan Dzakovic, a government whistle-blower and aviation security expert who worked for both the FAA and the TSA. Former American Airlines chairman Bob Crandall is characteristically blunt: "There isn't one goddamn politician with the balls to say what we're doing is silly. Because if he does and then an airplane blows up, someone is going to play it back to them." He adds, "It's all theater, that's all."

It should be noted that I was granted access to the top officials at the DOT, FAA, and NTSB, but my request for an interview with John Pistole, the administrator of the TSA, was not acted upon.

Playing Whack-A-Mole

Mention airline security and nearly all discussion is steered toward the kabuki of screening passengers in airports—the pat-downs, the X-rays, the confiscation of shampoo. But experts note that the most discussed point of entry is just one of many points of entry into airline security, and for more than a decade not enough research, focus, resources, or money has been channeled into other sectors that are potentially more dangerous than passenger checkpoints.

Prior to 9/11, the airlines themselves were charged with providing adequate security for commercial aviation. And in August 2001—just one month before the terrorist attacks—the Air Transport Association testified the industry was spending $1 billion annually on security. One month later the worst security breaches in U.S. history occurred, and the ATA revised its prior estimate and noted that . . . well . . . um . . . it's actually not $1 *billion* and is more like . . . $300 *million*.

"After 9/11 it became apparent the FAA had dropped the ball and the airlines were looking for the lowest-cost providers," says aviation law expert Paul Dempsey. "Security regulations have become Kafkaesque and well beyond the scope of what's necessary to prevent bad guys from getting on a plane." Before 9/11, carriers resisted every security measure, with one exception: airlines like that passengers must provide positive ID, since that has helped reduce the industry's problem of passengers reselling tickets.[2]

One would think the FAA would have beefed up surveillance of its vast database after the debacle of 9/11. But in July 2011 the inspector general of the Department of Homeland Security reported that the TSA had vetted about four million individuals licensed by the FAA and found 506 names that were "true matches" to the Terrorist Screening Database. What's worse, the TSA recommended that twenty-seven certificates be revoked. And there's little comfort in learning that the FAA has lost track of the ownership records on one out of every three aircraft in the United States. In 2010 the Associated Press reported that 119,000 U.S. airplanes—33 percent of the nationwide total of 357,000—had "questionable registration" records. And a U.S.-registered twin-engine Piper that crashed in Venezuela in 2008 with 1,500 pounds of cocaine on board: the only problem was the actual registration belonged to a similar Piper that was securely locked up in Washington State. Experts say the confusion over missing and falsified registration "tail numbers" can facilitate drug running and even terrorism.

Silencing the Best and the Brightest

Consider this quote from Bogdan Dzakovic in the *San Francisco Chronicle*: "The real problem is that the TSA is built on the same weak foundation as the FAA. It will always be at least one step behind the terrorists. Remember the shoe bomber? Right after that incident, the TSA made everyone take off their shoes at the screening checkpoints. Then we had the female Chechen terrorists who apparently hid explosives in their underwear. And so TSA screeners started groping female passengers until a big public outcry brought that silliness to a stop. Next time the terrorists might put explosives in toothpaste tubes, and you can count on TSA screeners squishing out all the toothpaste from passengers' bags." That appeared in print on July 9, 2006; exactly thirty-one days later, British authorities foiled the plot to detonate liquid explosives on ten transatlantic airline flights, and immediately thereafter the United States banned all liquids and gels, including toothpaste. In the weeks that followed, the restrictions were continually modified, tweaked, and revamped, and by September 26 the TSA's infamous "3-1-1" rule was instituted: "3.4 ounce containers in a 1 quart bag, 1 bag per passenger."

As for Dzakovic, he spent fourteen years working for the FAA, first as an air marshal and then as an elite "Red Team" leader who probes the system by posing as a bad guy and simulating attacks. After 9/11 he joined the TSA and almost immediately found the new agency had no desire to tap into the expertise he and other ex-FAA veterans possessed in abundance. When Dzakovic spoke up and spoke out, he was punished. "They put me in a job that's little more than clerical work," he says. "I could teach it to a high school kid in two hours."

What he is not teaching—and what he and others claim the TSA is not promoting—is how effective Red Team techniques can simulate the work of terrorists to make the entire network stronger and safer. Brian Sullivan, a retired FAA security specialist, is outspoken in pointing out the TSA is "a behemoth" that wastes billions of dollars: "If they took down multiple planes on any given day, that could push us over the edge. Is that still possible? And the answer is, yes it is."

Considering that the Department of Homeland Security—which oversees the TSA and twenty-one other government agencies—is the newest cabinet-level department, it has generated a disproportionate number of employee whistle-blowers since 2002. And the TSA has been charged with a disproportionate amount of malfeasance and waste. For example, a 2005 report from the inspector general of the DHS found that the rejection rate of applicants skyrocketed, from a ratio of 5:6 for hires-to-interviewees to 5:29. And corporate greed abounded: TSA's outsourced contractor, NCS Pearson, estimated its services at $104 million but was paid $741 million. In one case, the Wyndham Peaks Resort in Telluride, Colorado, was used to recruit fifty-one local screeners, and so billed the government $1.7 million.

As for the government/private sector revolving door that has existed at the DOT and FAA and NTSB for decades? The TSA caught up rather quickly. Michael Chertoff, tapped by the Bush administration as the second secretary of homeland security, left office in 2009 and formed the Chertoff Group. Although he would not reveal his client list, later that year it eventually came to light that it included Rapiscan, a company that manufactures full-body scanners, the same product he repeatedly hyped to the media. A few months earlier, the TSA had purchased 150 of these machines at a cost of $25 million. Passenger advocate Kate Hanni accused Chertoff of "shamelessly peddling his wares."

Still, the TSA has its defenders. MIT's Arnold Barnett asserts, "They really are showing a bit of courage and doing a fairly good job. They're under all sorts of pressure." He also says that security issues worry him much more than do safety issues: "It seems to me that if you find a lapse, you should try to take it off the table. It does seem monstrously reactive, but still, the more options you eliminate the harder it will be [for terrorists]." Barnett adds that he previously felt European airports were more secure than U.S. airports, but recently reversed that opinion.

Another thoughtful expert about security issues in general—and aviation security issues specifically—is author and speaker Bruce Schneier. He is at once critical of the TSA and quick to commend it. In a 2011 speech he declared that airport passenger screening is wasteful: "This is actually a stupid game and we should stop playing it. This is not security—it's security theater."

Schneier maintains that the actual track record since 9/11 has been amazingly good: Through 2011 there were no aviation security fatalities worldwide after August 2004, when two Russian airplanes were bombed simultaneously. And for U.S. airlines, there were two incidents since 9/11—the would-be shoe bomber and the would-be underwear bomber—and neither was successful. "I'd like to see more of us accept the mathematics of security," he has stated. "The most dangerous part of your airline flight is the taxi ride to the airport."

So what are we doing right? Schneier says the focus has been effective at the "beginning and end" of a flight, with intelligence and investigation in advance and emergency response at the scene. He says two positive developments since 9/11 are reinforced cockpit doors and passengers fighting back. As for what works at the airport, he cites three factors: the deterrence of passengers carrying weapons and bombs, the use of canines for detecting explosives (which presents logistical issues), and behavioral profiling (which also presents logistical issues).

So I asked him, what would he do if he were named administrator of the TSA? "The head of the TSA is the wrong place to address this," he said. "The TSA has too much budget already. It's not just about aviation. So it should be addressed at a level above the TSA." In other words, while we fret obsessively about airlines, we leave ourselves vulnerable to attacks against our subways, train lines, ports, bridges, tunnels, arenas, shopping malls, and other infrastructure.

Passenger Screening

For a decade now the TSA has been searching for just the right machines and methods to screen passengers and their carry-on baggage—but without success. What's more, there are equipment shortages, and little continuity of systems and technologies across the board, creating a patchwork quilt of screening processes, particularly at smaller airports.

In 2007 the U.S. Government Accountability Office stated:

"Concerns have arisen as to whether top management at [TSA] were negatively impacting the results of red team operations by leaking information to security screeners at the nation's airports in advance of covert testing operations." Subsequently, GAO's Forensic Audits and Special Investigations team conducted testing of its own and found that investigators passed through airport security checkpoints carrying prohibited explosive components without being caught by TSA security officers.[3]

On his site Ralph Nader writes at length about screening deficiencies. On the topic of radiation, Nader states: "Homeland Security should respond when physics professor Peter Rez of Arizona State University calculates the radiation dose to be ten times higher than [DHS] is asserting. Or when David J. Brenner of Columbia University's Center for Radiological Research says that using these scanners—with up to 1 billion whole-body X-ray scans per year in the U.S.—'may profoundly change the potential public health consequences to the population.'" Nader also asserts there are malfunctions, and cites John Sedat, "one of four scientists at the University of California at San Francisco who is questioning TSA's technical assertions." According to Nader, Sedat claims "these machines could stall, giving passengers 'severe burns if not worse'" since "software fails often."

The TSA has gotten high marks from some experts for finally introducing a Behavior Detection Officer program, which "utilizes non-intrusive behavior and analysis techniques to identify potentially high-risk passengers." Schneier, for one, is a strong proponent of profiling based on behavior—not race or gender—but he acknowledges it's difficult to implement proper training: "Profiling is a hard one because you want to avoid organized racism." This dovetails with the screening methods used by Tower Air's Israeli security teams on our flights to and from Tel Aviv; I used to watch as they screened reservations information weeks in advance and profiled passengers in the airport parking lot before they even entered the terminal.

Unfortunately, critics charge, the TSA is shortchanging its own efforts, through weak and sometimes virtually nonexistent training. "They heard the message, but their method is almost laughable," says Sullivan. "We have to get away from screening everybody. We have not made full use of dogs and behavior profiling . . . I think they're starting to recognize it's pure folly to pat down everyone. They're starting to look for bad people."

While many suggest more of an Israeli approach to passenger profiling, a recurring theme among U.S. security personnel is that one-size-fits-all doesn't work in the largest commercial aviation market in the world. As one American security expert says, "I get tired of the El Al people telling Americans what they should do, because they have one airport and forty planes. We have that many planes taking off *every minute* in this country."

I reached out to Barbara Peterson, who worked as a TSA screener and wrote about the experience for *Condé Nast Traveler* in 2007. Years later, she was floored by how little things have changed: "We still have no effective way to prescreen passengers and we're still treating everyone who transits an airport as a potential criminal." Working inside the system confirmed for her how pointless the security drill really is, without focusing much beyond the passenger checkpoint. Sadly, Peterson says she sees no sign of that changing.[4]

Of course, all this screening can add minutes or even hours to an air travel itinerary, and security is the prime source of the "hassle factor" cited by weary flyers. Shortly after its inception, the TSA introduced a terrific tool for consumers: an online Wait Time Calculator that allowed passengers to calculate how long they would be standing in security lines at specific terminals in all of the largest airports throughout the country. First the calculator was under construction, and then it disappeared.

Baggage and Cargo Screening

In July 2011 the GAO issued a report with a title that could have been used as an executive summary: "TSA Has Enhanced Its Explosives Detection Requirements for Checked Baggage, but Additional Screening Actions Are Needed." The GAO found that "some" explosives detection systems met standards established in 2005, while many others did not; what's more, "TSA has no plan in place outlining how it will approach these upgrades." Congressman Mica, the Florida Republican who heads the House Transportation Committee, is always eager to make the case for privatizing airport security, and he greeted the report by saying TSA continues to waste our limited resources and threatens transportation security.[5]

So the question is whether the TSA is more about jobs or about security. Blurring the issue is the internal debate within DHS over the exact responsibilities of screeners. The TSA boasts whenever an airport pat-down leads to an arrest for a nonsecurity issue, such as an outstanding warrant; others worry that undertrained, underpaid frontline federal employees are overstepping their bounds.

Transportation policy expert Stephen Van Beek underscores this when he says, "The TSA and CBP [U.S. Customs and Border Protection] have to be service oriented. They have evolved into law enforcement organizations rather than service-oriented organizations." He adds, "Sometimes it's just appalling the way passengers are treated." Appearances clearly are a factor. Despite its budgetary concerns, in its short history the TSA has found the money to revamp screeners' uniforms, and it's not lost on law enforcement veterans that the current model gives screeners the appearance of police officers—albeit without the proper training and authority.

Meanwhile, one of the best detection systems is decidedly low-tech: dogs. Experts say they still represent "the state of the art in real-time detection" of items of forensic interest, and they must be an integral part of interdiction or threat prevention systems.[6]

Most alarming of all is that experts say millions of tons of cargo and mail are carried on U.S. passenger airlines every day without being properly screened at all.

Federal Air Marshals

For years I have been speaking to air marshals and former air marshals who tell of the TSA's nonsensical methodology. And for years the most egregious policy debate focused on the strict suit-and-tie dress requirements, even on summertime flights to beach resorts; it became known as the "kill-me-first" dress code policy. They also speak of inflated employee ranks.

Robert MacLean is a veteran security expert who joined the Federal Air Marshal Service just two months after 9/11 and served until 2006; at that time he was "terminated against his will" for publicly reporting on a TSA internal directive to temporarily remove all marshals from long-haul domestic and international flights to address a budgetary shortfall by avoiding hotel bills. Although the information he shared with the press was considered "sensitive" and not "classified," his termination papers state: "Your unauthorized media appearance and unauthorized release of SSI information to the media raise serious doubts about your judgment and trustworthiness."

Many of MacLean's claims about nonsensical and downright dangerous policies in the TSA's early days were later supported by a GAO report in 2009, which cited mandatory dress codes and airline and hotel check-in policies that made undercover work all but impossible. A report from the Project on Government Oversight stated: "It is disturbing to learn how poorly the agency has handled federal air marshals' safety and security concerns."

Today MacLean's biggest concern is that the Federal Air Marshal Service has deviated from its primary mission and devolved into "jet bouncers." He emails me numerous news reports documenting air marshals embroiled in fistfights with drunks, hoodlums, and wackos, and says, "The bottom line is air marshals need to protect the flight deck. It's clear that if you douse yourself with booze and act like an idiot, it's an effective way to reveal the jet bouncers. If I want to disarm air marshals, I can set up a distraction and create a ruse and overwhelm them and then have a weapon." He and other security experts decry a recent laxity in hiring—many applicants are fresh out of college and no longer have military combat and/or police experience—thereby creating human "ammo pouches" for terrorists.

"Pistole needs to push the airlines to take more responsibility," says MacLean. Specifically, he believes airlines should hire their own security personnel for routine in-flight disturbances, and "deputized" passengers who provide assistance should be granted immunity from liability. Of course, in an industry infamous for cost cutting, there's no point in wondering how airline financial officers would view such a proposal.

Yet considering how many air marshals have been disciplined and fired for supposedly revealing classified data, the TSA itself is amazingly public in revealing sensitive information. The agency's site features a press release and photo of a marshal who assisted in an off-duty arrest. And in a truly bizarre bit of promotion a few years ago, the Sheraton Fort Lauderdale Airport Hotel actually issued an electronic flyer titled "Federal Air Marshall [*sic*] is Company of the Month in July!" The announcement included this: "Please feel free to spread the word!"

Cockpit Security

And then there's the cockpit. As Crandall says, "It's the most nonsensical thing to put pilots through security. If he wants the airplane to crash, he'll crash it. What kind of madness is this?"

Ralph Nader has been speaking out on airline security issues for more than forty years, when a wave of Cuban hijackings in the late 1960s highlighted vulnerabilities. But between 1968 and 2001 the airlines resisted his repeated calls to strengthen cockpit doors. One of the most immediate steps taken after 9/11 was that reinforced doors be installed on the entire U.S. commercial fleet. Amazingly, the FAA reported the mission complete by March 2002. But an investigation I conducted for *Consumer Reports* found fifty-one incidents between 2002 and 2007 with failures of the tougher doors: they unexpectedly opened in flight, the locks jammed, they broke when slammed.

And even if there were no problems with reinforced doors, experts note the inherent danger, particularly on long flights: eventually crew members need to open those doors for meals and lavatory breaks, which is why relying only on a reinforced door completely defeats the purpose. And one former air marshal's view of the common stopgap measure, a flight attendant blocking access with catering equipment? "Little Betty using a beverage cart isn't ready to brace against Al Qaeda."

Instead, many recommend a "secondary flight deck barrier," a concept endorsed by the Air Line Pilots Association in a 2007 white paper. In fact, Congressman Steve Israel, a New York Democrat, has repeatedly introduced legislation to mandate secondary barriers, without success. Here credit is due to United Airlines for voluntarily installing such barriers; the lightweight set of cables can be easily installed and disassembled by a flight attendant at the front of the cabin. But at an estimated twenty-five thousand dollars per aircraft, it's little wonder other airlines haven't emulated United.

Then there is the Federal Flight Deck Officer program, whereby marshals train pilots to use firearms as the last line of defense. But it has faced budgetary problems since its inception; in 2011 the Air Line Pilots Association reported the program had stopped accepting new applicants. Some security experts have expressed opposition, but the Airline Pilots Security Alliance makes a strong economic argument by stating it costs $700 million annually for FAMS to protect 2 percent of the nation's twenty-eight thousand daily flights, whereas it would cost $15 million annually for armed pilots to protect 100 percent of flights.

Who's Guarding the Planes?

A few years ago I obtained an internal ePassages memo sent by Northwest Airlines to its employees back in 2003; the update on maintenance outsourcing noted: "SASCO, one vendor used by Northwest, currently performs NWA work inside a Singapore military base. It is extremely difficult for a non-employee to gain access to this facility." Actually, it was odd Northwest should mention SASCO, let alone in the context of security. SASCO/ST Aerospace was an outside maintenance contractor in Singapore that served both Northwest and the U.S. Navy. One year *earlier*, in 2002, a SASCO/ST employee was arrested along with twelve others for their connections to a rather shady organization called Al Qaeda.

Imagine. Just months after the worst terrorist attacks in the nation's history, a major U.S. airline and a branch of the U.S. military were allowing their aircraft to be serviced overseas by an outsourcing company that had been infiltrated by the very same organization charged with killing nearly three thousand civilians on American soil. But we're not to worry?

There have been other security breaches at outsourced repair shops, even within our own borders. In 2005, U.S. Immigration and Customs Enforcement agents arrested twenty-seven illegal aliens working as aircraft mechanics for TIMCO in Greensboro, North Carolina, a facility whose client list included America West, Delta, and United. Subsequently it was revealed that once again the outsourcers were outsourcing, as several of the mechanics were seasonally subcontracted to TIMCO by a firm called SMART of Edgewater, Florida. In addition, one of those arrested was also charged with illegally purchasing a twelve-gauge shotgun. And the FAA's response? Kathleen Bergen, a spokeswoman based in the agency's Atlanta office, said, "To our knowledge, they all passed the written, oral, and practical tests."

Jack Blomfeld, who has worked in airline catering for thirty years, points out that outsourced service companies that provide aircraft handling, fueling, and catering have virtually unlimited access to airport tarmacs and commercial aircraft. But he also notes that the screening of such employees leaves gaping holes; in many cases the verification process doesn't even begin until after workers are on the job. "The cargo side of the airport is wide open," he notes. "You can stick twenty terrorists in a truck."

FAA and TSA officials don't want to consider it, but the offshoring of repairs—as detailed in chapter 7—raises security concerns as well. FAA whistle-blower Richard Wyeroski points out that U.S. commercial aircraft are particularly vulnerable when repairs are performed overseas: "They're worried about scaring the flying public, but I'd rather have them scared than dead. Our airlines should not be serviced overseas where we can't provide surveillance."

A mechanic for a major domestic carrier recalls that a few years ago a colleague removed a panel in the cabin of an aircraft that had just returned from a repair shop in Latin America and found two duffel bags of cocaine stored in the cooling and refrigeration units of the galley. An agent from the Drug Enforcement Administration told the mechanics, "It's not as rare as you would think." But the mechanic's concern, obviously, was that instead of cocaine that contraband could have been weapons or explosives. The airplanes themselves—even when serviced and parked at major airports—remain one of the most vulnerable links in the security chain.

The Privacy Debate

American privacy is more threatened now than ever, and the front lines in this battle are commercial airports. As *Popular Science* noted in 2003: " 'Outside of lab rats,' says industry consultant Michael Planey, 'airline passengers are the most analyzed subjects in the world.' "

Al Anolik, the travel attorney who has long been in the forefront of battling for passenger rights, may surprise some industry insiders with his view of security: "I am to the right of the spectrum on this," he says. "I am concerned about the threats. I have to give up some of my basic civil rights because of the threat to civil airlines. I don't like taking off my shoes, but it's necessary."

He argues that the funds have to be secured to provide the electronic equipment necessary to thoroughly screen passengers. "Travel is a privilege and not a right," says Anolik. "We know there is a threat. The U.S. cannot go into some of these issues. The captain has to have authority to act. And we need passenger profiling—it's a fact."

Not surprisingly, the American Civil Liberties Union disagrees; in 2010 the ACLU stated the government should enact procedures that pose "the least threat" to travelers' civil liberties but have also been proven to be effective. It further suggested that "routine" full-body scanning, "embarrassingly intimate" pat-downs, and racial profiling are what do not work. But finding that balance has proved elusive.[7]

11

Cloudy Skies: Aviation's Carbon Footprint

Fly between sea and sun!
Take the course along which I shall lead you.
—*Daedalus to Icarus*

IN 2011, WHEN CONSTRUCTION ON the busy 405 Freeway in Los Angeles so threatened to snarl traffic it was dubbed "Carmageddon," JetBlue stepped into the breach. The carrier offered four-dollar fares between Long Beach and Burbank (a distance of 37.3 miles according to MapQuest) and received "the cooperation" of the FAA to operate Airbus A320s at low altitudes. Clearly it was a silly promotion designed to do just that—promote JetBlue. But it also highlighted how airlines and government agencies are focused on corporate profitability even when it's at the expense of the planet.

Consider, for example, that unlike legislators and regulators in the United States, government officials in the European Union are actively seeking to address aviation's carbon footprint. Earlier this year, the EU introduced the first major program in the world that limits carbon emissions; airlines that fly within, to, or from Europe need either to reduce their CO_2, reduce flights, or purchase allowances. The goal is to cut 183 metric tons of CO_2 annually by 2020. Furthermore, even the industry's largest global trade organization, the International Air Transport Association, reported that the "net financial impact" may be slightly positive, or a little better than break-even.

Clearly that isn't good enough for U.S. airlines and the American politicians who support these corporations. Last year Congressman John Mica, the Florida Republican who chairs the House Transportation Committee, held bipartisan hearings to fight the EU policy. Airlines for America, which represents domestic airlines, claimed the "scheme is illegal" and filed suit, along with American and United/Continental. But the Environmental Defense Fund and other groups were quick to question such congressional motives. "It's simply baffling that these legislators are working so hard to keep U.S. airlines in the dark ages of relying on inefficient airplanes and outdated technologies," stated Annie Petsonk, the EDF's international counsel.

Legal scholar Paul Dempsey, a vocal critical of airline deregulation, points out that aviation is the only industry that pollutes at high altitudes and we have a very poor understanding of what that means. He adds that airline codesharing and marketing deals pose tremendous threats because "we end up putting too many seats in the skies," without giving a moment's thought to emissions consequences.[1]

I turned to Dudley Curtis, the communications manager for the European Federation for Transport & Environment, an environmental organization that campaigns specifically on transportation issues. When I asked him what airlines should be doing, he responded: "As a first step they should be supporting, not attacking, pretty much every conceivable policy initiative worldwide to address the environmental impact of aviation." I have come to agree with him, and conclude the industry's behavior is in keeping with its blindly striking out at pro-consumer initiatives.

Defining the Dangers from Above

The airline industry's environmental footprint takes many forms, through myriad means such as land contamination, aircraft deicing disposal, and exhaust from large fleets of diesel-powered airport ramp vehicles. Then there is noise pollution. But nothing compares to the dangers posed by carbon emissions, particularly at high altitudes. This was even acknowledged by the FAAC's Environment Subcommittee, chaired by Bryan Bedford, CEO of Republic Airways.

It seems ironic that the CEO of two regional carriers was tapped by the government to address environmental concerns, since one could argue that sending so many smaller planes into the air poses a heightened risk to climate change. Thankfully, the subcommittee also included Juan Alonso of Stanford University's Department of Aeronautics and Astronautics. Shortly after I met him, I joked that if there was one person on the FAAC lonelier than the lone consumer advocate, it was the lone environmentalist.

The good news is Alonso thinks there's a new awareness of green aviation, due to its relative importance and the social costs. He asserts that since 2003 there has been focused research on fact-based, independent research opinions. Despite his advanced degrees and the complex nature of his work, Alonso has a knack for explaining CO_2 emissions rather simply: in goes fuel, out come emissions. The only answers are using less fuel or finding alternative fuels. And technologically, there are solutions.[2]

The best minds believe aviation is directly responsible for about 2.5 percent of climate change; however, these calculations could be off by a factor as high as two, so 5 percent is not an insignificant measurement. What's more, aviation is solely responsible for many forms of high-altitude emissions. And according to the International Civil Aviation Organization, the amount of CO_2 emissions from aviation is expected to grow around 3–4 percent per year.

What is not in dispute is that a variety of aircraft emissions are affecting climate, including water vapor, soot, and CO_2. Low-level emissions in the range of 1,500 to 3,000 feet affect local areas for a radius of about ten miles, and high-altitude emissions can last three to five years. These include small particles that present some of the most serious health concerns.

The dangers keep accruing, in the form of nitrogen oxide, sulfur, and methane. As Alonso tells it, the science can be sobering, since CO_2 lives in the atmosphere for up to three hundred years, regardless of its location. In addition, experts worry about high-altitude aircraft water vapor, which can block the sun and have a significant climate effect.

There are other dangers as well. Experts note that we shouldn't forget air quality, since aviation's impact on air quality is significant. In fact, it's estimated there are fifty thousand deaths annually in the United States and one million worldwide due to poor air quality, which remains a significant public health issue.[3]

Ian Waitz, dean of the School of Engineering and a professor of aeronautics and astronautics at MIT, acknowledges that airplanes have a high-altitude impact that is "perhaps double" the effect of CO_2 alone, and this problem is primarily manifested through contrails and cirrus clouds. But he adds, "The good news is some relatively modest changes could alleviate this." Theoretically, airlines could modify flight plans en route by altering altitudes and airways to lessen the impact of contrails and cirrus. Although this could lead to higher overall emissions, Waitz believes that within the next five years we may elect to trade greater CO_2 output to diminish these more harmful effects.

But Billy Glover, vice president of environment and aviation policy for Boeing, points out that addressing the contrail issue by altering flight patterns and altitudes translates into the potential for more CO_2. He adds, "If we had a solution for contrails, we would get benefits immediately. If we had a solution for CO_2, the benefits would not be seen for one hundred years."

Alternative Energy Sources

Much progress has been made when it comes to biofuels. Some reasons for optimism in recent years include Virgin Atlantic flying a Boeing 747 on a mix of coconut oil and other biofuels, KLM operating a scheduled flight with a Boeing 737-800 on bio-kerosene, and Lufthansa announcing daily commercial flights with an Airbus A321 utilizing a biofuel blend of 50 percent hydroprocessed esters and fatty acids. (It's worth noting these achievements were accomplished by non-U.S. airlines.) Even so, not everyone is impressed with such advancements, under the "too little, too late" theory. But for the short term at least, alternative fuels are raising hopes in some quarters.

It's a topic being discussed at the highest levels of government. At Georgetown University in March 2011, President Obama spoke on "America's Energy Security" and stated, "I'm directing the Navy and the Departments of Energy and Agriculture to work with the private sector to create advanced biofuels that can power not just fighter jets, but also trucks and commercial airliners." In July 2011, the U.S. Department of Agriculture announced a joint agreement with Seattle-based AltAir Fuels to produce sustainable crops in California, Montana, and Washington. But when I asked Secretary LaHood about alternative fuels, he replied, "I think it will be left to the airline industry. And I think they'll be looking at those opportunities as they buy new planes."

Unfortunately, the domestic airline industry has not been in the forefront of developing such new technologies. But one airline executive has: Sir Richard Branson, CEO of British-based Virgin Atlantic Airways, who has invested billions into the Virgin Green Fund. As he told *Time*: "There's a frightening potential scenario out there that means that anybody who's in a position to do something must do something. In particular because I'm in one of the dirty businesses, the airline business, I've got all the more responsibility to do something."

Waitz maintains that so much these days depends on government incentives, which can lead to further development of algae or other crops as biofuels. But he stresses these are scarce resources, and we may find better uses for them than aviation: "Sometimes the aviation industry looks at these issues only from an aviation industry standpoint."

He is not alone. Many environmentalists worry that the "food-to-fuel" conundrum will cause more harm than good for a planet with starving people. That's why Waitz fears biofuels for airplanes may be mandated, because that could develop into a costly and inefficient use of a scarce resource. For example, he doubts ethanol will be a successful alternative to jet fuel because of its expensive supply chain. Waitz adds, "The question is, will biofuels be lower cost at the point of delivery?"

I asked Curtis if biofuels are a cause for optimism, and he responded: "No. Most biofuels sold today are worse than fossil fuels in terms of total carbon emissions, particularly when indirect land use change is taken into account [emissions associated with land clearance, often in tropical regions, to grow the displaced food crops]. The industry is making a lot of noise about biofuels, but it is far from making any significant deployment. It's a similar tactic to the American car industry who spent years lobbying against tightened fuel economy standards and instead lobbied for 'flex fuel' credits for cars that could run on corn ethanol, which is environmentally dubious in any case." That's a sobering conclusion, and it's worthy of a big-picture examination by the U.S. government—assuming Corporate America doesn't derail such a discussion.

The High Cost of Oil: Good News?

These dark clouds may literally be infused with a bright lining. Because even though airline executives may not be motivated to address the industry's carbon footprint, the rising cost of oil has motivated them to seek more fuel-efficient aircraft and engines. When Airbus began developing the superjumbo A380, airlines bidding on the project were doing so when oil cost $25 to $30 per barrel—not more than $100 per barrel. This disparity translates into about 3 percent of the plane's costs. As Alonso says, "That's the difference between being the most profitable airline and going bankrupt."

While recognizing that fuel remains the most volatile cost for airlines, analyst Helane Becker also notes the industry will have to address its aging aircraft problem. In the late 1970s and early 1980s, three-engine McDonnell Douglas DC-10s and Lockheed L-1011s and four-engine Boeing 747s gave way to a new generation of two-engine jets, including twin-aisle wide-bodies like the Boeing 767 and Airbus A300, which were among the first certified to cross oceans with fewer than three engines. At the same time, three-person cockpits evolved into two-person cockpits, as flight engineers went the way of buggy whip makers. Losing one crew member and one or two engines were huge economic advancements for the airlines, producing savings in two accounting sweet spots, fuel and labor. Becker believes the next "game-changer" technologically will not necessarily be a new airplane, but new biofuels: "From an airline equity standpoint, I think the biggest issue is higher energy costs."

One would think an industry so reliant upon a resource as scarce as oil would be encouraged by investors to seek alternative forms of energy, particularly since it represents 35 to 40 percent of an airline's operating costs. Amazingly, the nation's small army of airline analysts almost never mentions environmental concerns—with the notable exception of Bob Mann of R.W. Mann & Company, who notes there is a "practical side" to worrying about the planet's future.

Reducing fuel consumption obviously remains a high priority, but when I suggested the industry has maxed out its ability to make lighter commercial planes, Alonso warned me not to sell short such initiatives. He explained that three key factors remain: reducing aircraft weight, improving the aerodynamics of aircraft, and improving fuel consumption of engines. He noted, "The vast majority [of progress] will come from engine efficiency." Alonso added that the jet engine industry has been "much more aggressive" than the airframe industry, particularly in Europe. What's more, he suggested the aircraft industry is a bit out of step in pushing extra capacity and extra range, when a leading customer such as Lufthansa is instead demanding better cost efficiencies.

I asked Waitz if aircraft and engine technology has plateaued in the pursuit of reducing carbon footprints, and he said, "It certainly has gotten harder and harder to make change. But you can do more. The question is, at what cost?" He added, "It's tremendously challenging technically and financially to introduce a new airplane."

But experts say there are other options to reduce airline emissions, and some ideas are unfathomable. Among them are flying slower; eradicating first class, and employing "cattle car" seating; removing all cabin windows; and eliminating cockpit crews and replacing them with full-time autopiloting systems. Clearly such suggestions raise serious issues about safety, comfort, and convenience. That's why Waitz asks, "What level of cost or risk are you willing to assume? You need to find the balance between a broad set of society's needs."

These are tough problems, and unfortunately the solutions are not easy, either. Medium-term mitigation for CO_2 emissions could possibly come from improved fuel efficiency. However, this will only partially offset the growth of CO_2 aviation emissions.

Next-Generation Air Traffic Control?

In an editorial for CNN.com in April 2011, Southwest Airlines CEO Gary Kelly stated: "It's not an exaggeration to say that today's air traffic control system is using 1950s technology and flight paths to route our aircraft during a time when most drivers on the highway are following direct routes guided by their GPS systems."

If there was one issue every member of the FAAC agreed on, it was the necessity for modernizing the nation's air traffic control (ATC) network through the Next Generation Air Transportation System, known as NextGen. It also was the only topic discussed on all five subcommittees, since it affects safety, competition, finance, labor, and certainly environment. From a consumer perspective, NextGen should mean safer operations, less congestion, fewer delays, and improved service.

So what is NextGen? Rather than a new Univac computer that gets plugged in to the accompaniment of a ribbon-cutting, the system has countless moving parts and components. It involves gradual rollouts of GPS technology on the ground and onboard aircraft, coupled with a switch to data communications and upgrades to all the many existing technologies that oversee the nation's civil and military air traffic.

Secretary LaHood is enthusiastic about NextGen. He believes it will help relieve congestion, so air traffic controllers can guide planes in and out of airports more directly. Former FAA administrator Randy Babbitt agreed: "This is truly an investment." He maintained that once NextGen is fully deployed by 2025, the savings will be in the multiple billions. Conversely, *not* adopting NextGen will cost the American economy up to $22 billion a year.

So there's nothing not to like about NextGen. Except, of course, the obvious: Who's going to pay for it? Congress, the DOT, and the airlines have engaged in an odd kabuki for years over who will pay for what, and when. It's difficult to directly monetize the benefits, so there's a hesitancy to step forward and invest.

As for funding, LaHood acknowledges the challenges: "We have a responsibility as the federal government, frankly, to be the big players in this. I think we do have a big responsibility. We're pushing this. . . . But the lion's share of the funding should come from the federal government and we know we'll have partners out there who will help us. We know the airlines are limited on their resources although they're very supportive of NextGen."

In that editorial Kelly committed to Southwest's investing in NextGen technology, and also wrote: "But the airlines cannot reach NextGen alone. The FAA, in coordination with the aviation community, must focus its limited resources to design and implement GPS-based flight paths that will result in measurable reductions in fuel consumption and emissions. In other words, it does no good to use new technology to fly the same old routes more precisely. We must have new flight procedures approved by the FAA to leverage the tremendous potential of NextGen technologies."

However, the next big push from the airlines will be for further privatization, not just of TSA screeners but the FAA employees who staff the nation's control towers. Flight Safety Foundation's Bill Voss says, "I'd trade the free market pressures any day over congressional oversight. There are a lot of earmarks and pork in the system. The trouble is this is funded by user fees, and that's a toxic issue."

The problem is that in aviation the track record shows that free enterprise unfettered by government oversight has led to cutting corners. And with air traffic control, the risks extend beyond the obvious—the safety of the public, both in the air and on the ground. They also extend to the future of the planet.

Fighting the Not-So-Good Fight

"What are the economic incentives for the best interests of the whole country?" asks transportation policy expert Stephen Van Beek. "We need to look at how to transition to greener transportation." He notes how airlines buy and sell airport slots, a practice he suggests should be stopped immediately: "The airspace is owned by the public. Some bank out there is giving money to an airline for assets that are not theirs." And Van Beek points to the Los Angeles–San Diego market, which is served with twenty-seven flights daily by United Express—all with thirty- to fifty-seat aircraft: "From a system point of view, that doesn't make sense."

For these and other reasons, airlines are favorite targets of environmentalists. One common complaint is that the industry often engages in the practice of "greenwashing," or touting environmental programs through its marketing and advertising campaigns. In fact, the Angry Mermaid Award—which is bestowed by a coalition of environmental organizations based throughout Europe and the Far East—was established "to recognize the perverse role of corporate lobbyists, and highlight those business groups and companies that have made the greatest effort to sabotage the climate talks, and other climate measures, while promoting often profitable, false solutions." A recent nominee was the International Air Transport Association, which was "honored" for a decade of "leading the industry's efforts against regulatory action on climate change." Among the issues the citation referenced were fighting for aviation to be exempt from green regulation; lobbying against aviation to be included in the UN Copenhagen climate change conference in 2009; and referring solely to CO_2 emissions and ignoring the impact of nitrogen oxide emissions, contrails, and cirrus clouds, which have impacts "two to five times greater" than CO_2 alone.

When it comes to greenwashing, the airlines aren't alone. It seems fair to worry that we're nibbling at the edges of this issue while not focusing enough on the elephantine impact of jet engine emissions. For example, Greenopia annually ranks North American carriers using a broad array of criteria to determine which are more environmentally friendly; in the 2011 results, Virgin America led ten U.S. and Canadian airlines, while American ranked last. Four green leaves for Virgin America and only one green leaf for American would seem to indicate a huge difference in the carbon footprints of these carriers, no? Yet a report by TreeHugger.com reveals that the criteria include fleet age, fuel consumption practices, carbon offsets, green building design, recycling programs, and green food items available in-flight.[4] Jeff Nield, the author of the TreeHugger.com piece, who suggests "green airline" is itself oxymoronic, makes a strong point: "When it comes right down to it, the fact that airplanes spew forth fossil fuel emissions up in the atmosphere is what makes them such heavy polluters." Put another way, catering with fair-trade food items is a fine practice, but it won't reduce high-altitude emissions.

For years now, green groups have been debunking the airline industry's claims about its collective carbon footprint. In the forefront of this charge has been Transport & Environment, which receives funding support from the European Commission. The claims are that the current industry strategy is to talk about "miracle" technologies and what other entities—such as governments—should do to address aviation's footprint. T&E also believes the airlines are interested only in unenforceable, nonbinding voluntary commitments on reducing emissions.

Dudley Curtis is not generous in his assessment of the United Nations–backed ICAO ("It has failed miserably to fulfill its obligation to cut aviation emissions under the Kyoto protocol") or the industry trade group IATA ("It proudly touts its environment record, which so far can be summed up as 'business as usual' with a couple of nonbinding, voluntary, unenforceable, and in any case weak commitments based on current fleet renewal trends").

As for aircraft and engine manufacturers, Curtis says they should give their full backing to a meaningful global fuel economy standard for aircraft, one that reflects fuel use across the entire flight, not just when cruising at a fixed speed. Also, they should design planes for fuel efficiency rather than for increased range and speed, since fuel efficiency has stagnated over the last decade. Curtis asserts that the average aircraft flying today is no more fuel efficient than the Lockheed Super Constellation, the last generation of propeller passenger planes of the 1950s.[5] I asked Boeing's Glover about this, and he responded, "We have design requirements that include thousands of things. They're all folded in. We set our requirements in a very aggressive manner."

The airlines' deep pockets undoubtedly have kept regulators and legislators from demanding more from the industry. Because of the industry's lobbying, Alonso believes Congress will never enact a reasonable energy policy: "Unfortunately, they're going to respond when it's too late on climate change, but it certainly won't be in the next five years. Are they waiting for the first floods?"

Yet Alonso remains optimistic that the problems can be solved. He believes efficiencies can reduce current fuel consumption by as much as 80 percent. ICAO and IATA are not quite as optimistic but do assert that by 2050 the airline industry will produce CO_2 at 50 percent of 2005 levels. Alonso sums it up: "I do think suboptimization is better than no optimization at all."

Waitz believes Washington must do more: "It's hard to tell the private sector to invest. It's really the government's role to put in place incentives. We ought to have programs in place. Broad, flexible programs such as cap and trade. They are popular in Europe—but not in Washington." He notes that whether it's aviation regulations in Europe or automotive regulations in California, some communities have decided on more stringent rules and as a result they're driving innovation.

The airline business may be something of a punching bag, according to Waitz, because of its high profile (no pun intended). He claims, "People say, 'Look at this big dirty industry.' But this industry has high abatement costs." He points out that when assessing the carbon footprints of various industries, some sectors may do less and some may not even address such problems at all, but there are different views of equity in determining fairness. Waitz is optimistic as well; he says, "I think we will see continued innovation and continued reduction in carbon footprints. But not at the same rate as other industries. . . . You have to change your view of what's fair. There will be more pressure on aviation than in other industries."

Consumers have a role, too. We vote not only with our ballots but also with our wallets: airlines will need to pass costs on to passengers and therefore are missing opportunities to reduce energy costs, because unfortunately the airlines don't commit to enough research.

Before consumers can become more involved, however, it would help if they were given accurate tools to measure their own carbon footprints. Last year a friend proudly told me she was saving the planet by driving her Prius on a long trip to Washington, D.C., from New Jersey. But Alonso points out that a smelly yet crowded bus might actually be more eco-friendly. In 2008 I researched carbon calculators on behalf of *Consumer Reports* and quickly found they are inaccurate to the point of absurdity. Alonso agrees there's a ripe opportunity here, for accurate consumer calculations. One study found that a fully loaded Boeing 787 flying from San Francisco to New York City offers fuel economy similar to that of a Honda Accord with three passengers, at almost ten times the speed.[6]

As for carbon offsets, I suggested to Alonso they are reminiscent of the ancient Catholic Church practice of buying indulgences, so that rich sinners can keep on sinning. But he had a different view: "At the same time, people who incur the costs should pay for it. It's about individual responsibility, so you can at least write a check. . . . If it has a positive impact, then I'm fine with it."

But Alonso also pointed to the morality of the "inequity issue" inherent in air travel. Overwhelmingly, aviation emissions are produced by developed nations, but their effects are felt throughout the world, including underdeveloped nations. As he said, "The rich pollute, and the poor pay."

12

Ink-Stained Wreckage: How Airlines Manipulate the Media

> He [Adolf Hitler] is undoubtedly a great man, and I believe has done much for the German people.
>
> —*Colonel Charles A. Lindbergh, in a letter home from a 1936 inspection tour of the Luftwaffe*

AT ABOUT 10:35 ON THE morning of December 17, 1903, for the first time in the history of the species a human being took to the skies in a heavier-than-air machine that moved forward under its own power. Later that day, mere hours after first Orville Wright and then Wilbur Wright established and promptly smashed new altitude records, a working reporter managed to commit the first major screw-up in the history of aviation journalism. *Nota bene:* it would not be the last.

Even today it's not an overstatement to rank that first flight as one of the ten most significant achievements in the history of *Homo sapiens*. In 1903, though, it should have ranked in the holy trinity of journalism alongside the headlines SOME HAIRY GUY INVENTS FIRE and SOME OTHER HAIRY GUY INVENTS WHEEL. So you can imagine Wilbur and Orville's surprise when they decided to grant a world exclusive to the Associated Press—and the AP immediately passed up the story.[1]

The Wrights were good kids and the sons of an Episcopal bishop, so it's doubtful they said what they should have said. Namely . . . *WTF??* The longest of the four flights in Kitty Hawk, North Carolina, that day had lasted fifty-nine seconds but it got mangled as fifty-seven in the telegram sent to a third Wright brother, Lorin. No matter. According to *The Bishop's Boys: A Life of Wilbur and Orville Wright* by Tom Crouch, Lorin personally walked the news over to the offices of the *Dayton Journal*, the brothers' hometown rag, and handed it to Frank Tunison, the local rep for AP. The reporter's response is now legendary among those who track Media Reports That Never Were. "Fifty-seven seconds, hey?" Tunison barked. "If it had been fifty-seven minutes then it might have been a news item." Meanwhile, an editor in Norfolk, Virginia, had heard about the flights in nearby Kitty Hawk. He wrote up a somewhat muddled account of the event and contacted twenty-one newspapers to see if they would like to reprint it. Only five even bothered to respond.

Thus began the bumpy relationship between powered flight and the media that continues to this day.

Ninety-six years later, in February 2000, JetBlue Airways launched service from JFK International Airport to Buffalo and the *New York Times* reported that the company's inaugural departure was three hours late. Airline executives and others who were present swore the delay was not nearly so long, and also noticed that the *Times* reporter hadn't even been at JFK. The byline? Jayson Blair, the plagiarist later exposed for reporting from locations he had never visited, when the *Times* famously fired him in 2003.

Covering the Beast: Reporting on a Complex Business

As a journalist, I've written about the airline industry for many kinds of readers: airline executives, travel agents, corporate travel managers, investors, business travelers, and everyday customers. Like one of the proverbial blind men grasping at an elephant, each of these audiences is concerned with a different aspect of the massive beast known as commercial aviation. And what a beast it is. In 2009, the FAA reported commercial aviation was ultimately responsible for 5.6 percent of gross domestic product. It also helped generate $1.3 trillion in economic activity and 12 million jobs. And imagine the impact on our nation's economy if the airline industry weren't so intent on offshoring everything from telephone reservations to major aircraft maintenance.

There are quite a few excellent aviation journalists practicing their craft today, and I've been proud to know and even work with many of them. But a business as large as the airline industry inevitably is covered by a broad range of media outlets, and every day there's a tremendous amount of misinformation churned out. Kevin Mitchell battles for passenger rights as chairman of the Business Travel Coalition, and on many occasions he has found himself fighting long odds, because the lobbying efforts of Airlines for America and the individual airlines are so formidable and have the power to influence Congress, the DOT, and the media. Like special ops psychological warfare, these campaigns address why the airline industry gets so much coverage that favors it.[2]

As further example, Mitchell cites the period immediately following 9/11 when the airlines took advantage of contract clauses to lay off thousands of employees and blamed it on terrorism, and such claims went virtually unchallenged in the press. Labor leader Pat Friend underscores this point: "9/11 did cover a lot of sins." She notes that on the day of the terrorist attacks, there were 22,000 flight attendants at United, and ten years later there were 15,000.

And yet despite the full-court media press, the airlines do not enjoy a favorable reputation, which makes one wonder how poorly they would be perceived without spending all those millions. A longtime aviation journalist spells it out: "I think the airlines are their own worst enemies. They really have such a bad image. They're up there with polluters and oil companies. . . . A lot can be traced back to the historical reality that they are public utilities."

There are deeper issues than just bias here. Much of what is reported about the airlines is rather superficial, considering what is at stake. Travel columnist Joe Brancatelli points to the legalese embedded in the airlines' contracts of carriage and notes, "The airlines say, 'We don't guarantee we'll provide the flight.' No other industry says this about their products. And then the airlines want to know why people don't like them. Where is the consumer media explaining what is going on?"

Then there is safety. Consider that in 2010 *Time* published "20 Reasons to Hate the Airlines: A Brief History of the Industry's 30-Year Campaign to Nickel-and-Dime Us Nearly to Death," which chronicled such grievances as paying for headsets, paying for sodas, and paying for blankets. Annoying? No question. But for a headline invoking such strong terms as *hate* and *death*, it's odd the most dangerous condition cited is "The Disappearance of Legroom." None of the many safety, security, and outsourcing issues cited in this book even cracked the top twenty.

For years now some of us have been writing about critical life-and-death safety issues, such as the FAA's oversight of aircraft maintenance and regional airlines. But there's been no momentum, no steady drumbeat, as major media outlets have provided infrequent or nonexistent coverage of threats to airline safety. As Kevin Mitchell notes, "One of the reasons people are pissed off about three-dollar Cokes [on flights] is because fees have gotten so much more ink than outsourced maintenance. It hasn't gotten the visibility."

So why is that? Is coverage of airline safety too gory or depressing? That hardly seems plausible, in a culture that provides media saturation of natural disasters and violent crimes. Is it bad for business? Perhaps. But you would need a software programmer to decipher which media conglomerate is funded by which sister companies and outside investors. The likeliest reason seems to be the tremendous economic pressures on traditional media outlets, as they scramble to do more with less, and increasingly depend on less experienced and *outsourced* employees.

Searching for the Black Box

I have a simple rule when reading about aviation: I see if the author has spelled Delta Air Lines correctly (Delta is one of the last carriers to still break out *airline* into two words). If it's written as *Delta Airlines*, then I question every other premise. The same if the author employs such terms as *direct flight* (in airline parlance that's actually an *indirect* flight since it stops en route; the term is *nonstop* flight); *stewardess* (anachronistic and sexist); and *black box* (actually, there are two boxes, and they're orange, not black). I once stopped reading an airline story because it referred to "LaGuardia International Airport" in referencing what is a domestic facility.

They say we often fight the last war. Well, reporters covering airplane crashes are often covering the last crash, even when common sense dictates this one wasn't due to wind shear or terrorism or a short runway. This syndrome reached its peak in 1996, when TWA 800 departed from New York City and then exploded in midair off Long Island. I heard a local TV reporter suddenly speculate: "There's no word yet on whether this might have been attributed to faulty deicing." An old colleague from Pan Am called and let loose with the obscenities: "De-effing-icing! In July! Ninety effing degrees Fahrenheit! De-effing-icing!"

At other times, a lack of information can seem worse than misinformation. Joe Brancatelli recalls Northwest Airlines using extreme measures to thwart press coverage during the strike by its mechanics in 2005. This included banning reporters on company property and blacking out flight display monitors in airport terminals (obviously to avoid images of FLIGHT CANCELED messages). But the company left a hole—the flight tracker on its website remained in service.

No other journalist in the country did a better job than Brancatelli did covering that strike. Every day, several times a day, he posted updates on flight delays and cancellations, putting the lie to Northwest's claims that all was fine. ("Pay no attention to that man behind the flight information display screen!") Then he raised the stakes even higher by providing detailed coverage of the maintenance problems the FAA said weren't occurring at Northwest: the cancellations and air turnbacks and emergency landings. A few days later, Northwest put out the real flight cancellation numbers. For Brancatelli, the lesson of the Northwest strike remains an optimistic one: "Eventually, in an industry like this, the truth will come out."

Business Journalism: Ignoring the Rest of Us

And then there's "business news." If you examine coverage of the airlines you'll soon realize that the focus is almost entirely on investors—not customers, workers, or communities. We as a nation have not addressed how the best interests of a tiny handful of billionaires diverge from the best interests of the other 300 million of us, but the bias soon becomes apparent. Most reports are "balanced" only by the illusion of canned statements produced by a given company's public relations machine, peppered with "analysis" provided by tainted parties, namely Wall Street investors who stand to benefit by offering their prejudiced comments.

Sometimes outside opinions are shut out entirely. In May 2000, when the airline merger rumor du jour was a United Airlines–US Airways coupling, a disturbing story came to light: a publicist hired by the two carriers offered three major newspapers a sweet deal—with strings attached. In exchange for exclusive details on the $5 billion merger, the reporters promised not to make any calls for outside comments. The *Wall Street Journal*, *Washington Post*, and *New York Times* all agreed, despite the obvious danger of serving as the airlines' public relations arms.[3]

When the audience is the business community, the challenge for journalists is to provide alternative viewpoints on airline issues—from labor, communities or cities, and certainly consumers. On the eve of that Northwest strike, the *New York Times* did a good job of addressing the big picture, including the perspectives of passengers and employees. That was critical, considering that one report noted, "Clearly, Wall Street is rooting for a walkout at Northwest." That was no idle assertion but in fact was backed up by a quote from Jamie Baker, an airline analyst at JP Morgan, who stated: "Frankly, we're hoping for a strike." Imagine. An airline strike is a debilitating, demoralizing, and heartbreaking saga, and in the case of a maintenance strike, safety is undoubtedly at risk, yet Wall Street hoped for one nonetheless.

While filming *The Red Tail*, which documented the Northwest strike, Minnesota filmmaker Dawn Mikkelson was struck by how the media responded: "It was lazy coverage, in my opinion. It was, let's talk to economic analysts and let's talk to passengers. They never touched on the outsourcing issue. This was about cutting costs."

The airline industry/Wall Street/media axis can be a hall of mirrors at times. Many of the leading financial analysts and consultants quoted in aviation reporting have vested interests in the very issues they discuss—an article about airline consolidation, for example, may include puffy comments from a Wall Street analyst being paid merger-and-acquisition fees by the industry. But there are best practices here as well. When I requested an interview from Helane Becker at Dahlman Rose, her financial analyst firm, she forwarded a detailed seven-page disclosure statement for both investors and journalists that listed all the airline companies she rated, as well as her firm's relationships (if any) with those airlines. Kudos to Becker, but unfortunately not enough industry analysts emulate such practices, and not enough journalists ask before their interviews.

Spoon-Fed News

Reliance on public relations juggernauts is crippling American journalism. Like the "news" reports that breathlessly recite the weekend's box office receipts, or Black Friday retail revenues, or Super Bowl television commercial ratings, this is all lazy and superficial reporting. These days it's entirely possible to write and publish a news story about an airline bankruptcy filing without communicating *in any way* with a single human being. Cut and paste airline executives' quotes from the carrier's press release. Cut and paste quotes from Wall Street analysts' briefings. And if there's space, cut and paste a brief quote from a labor union's release. Serve warm.

Brett Snyder—better known as the Cranky Flier of CrankyFlier.com—illustrates the challenge that is presented to editors, producers, and bloggers every day: some airlines are more than happy to provide ready-made editorial content. He explains: "There's no question a good public relations or marketing firm will not pay for coverage. But there are bad PR and marketing firms that will. I'm approached all the time. They want you to pimp your product."

And sometimes press releases are just wrong. In May 2011, Cheapflights.com issued a release titled "How to Save on Hidden Travel Fees," which included the following: "If you're traveling with family, evaluate if you want to buy a seat for a child. Most airlines don't charge for children under the age of two to fly, so weigh if you want to hold your child or use an arguably safer car seat on the flight." Arguably? There is no argument—none whatsoever—among aviation safety experts about whether a child in a restraint system is safer than a lap child. Unfortunately, too many news organizations reprint such releases without any form of fact-checking.

And even the best reporters can be duped by industry spinmeisters. Consider the *CBS Evening News with Katie Couric*, which in 2010 reported on the DOT's new stranded-passenger policy, but either intentionally or unintentionally spun it in favor of the airlines: "New Tarmac Delay Rules Could Up Cancellations." This piece reflected the airlines' unfounded and ultimately untrue claim that the new policy, which requires that passengers trapped on the tarmac be freed from taxiing aircraft after three hours, would lead to more flight delays. Strangely, the segment included a quote from the CEO of Hong Kong–based Cathay Pacific Airways, a very odd source for a story detailing what was at the time a domestic policy. And it also included an unattributed omniscient statement: "Now the airlines say instead of delaying flights, they may just cancel them outright."

Yet this entire issue amounted to airline industry blustering and bullying. The proof came by October—just six months later—when DOT monthly statistics confirmed that *no* domestic flights were delayed for three hours or more, and flight cancellations *fell.* The DOT was proven right, while the airlines and their water carriers were proven to be needlessly crying wolf, but this clearly demonstrated how the powerful airline lobby can steer a public debate.

There's yet another form of media bias that permeates aviation coverage in the United States (and around the world). Companies and events with high profiles in media capitals receive a disproportionate amount of coverage, and the inverse is true as well. A few years ago a producer for one of the major morning talk shows was briefing me in the greenroom about my segment on airline baggage fees. While discussing which airlines charged the most, I pointed out that in fairness we should mention the two domestic carriers that didn't charge at all for first checked bags, Southwest and JetBlue. His nose crinkled and he said, "You can mention JetBlue, but not too many people have ever heard of Southwest." Spoken like a true New Yorker. When I politely pointed out to him that in terms of domestic passengers carried, Southwest was *the largest airline in the United States*—and about five times the size of JetBlue—his facial expression made it clear he thought I was either lying or wrong, and he suggested we just skip it.

Of course, JetBlue is based in New York City, and Dallas-based Southwest had only entered the Big Apple market in 2009. Conversely, JetBlue has both lived and died by operating in the glare of so much media coverage. Compared to other start-up airlines, it received an unusually high percentage of exposure when it launched in 2000; one former employee acknowledges it was the "Teflon airline." On the flip side, it withered under the intense coverage of its infamous "Valentine's Day Massacre" operational meltdown in 2007; I'm certain that a carrier based in St. Louis or Milwaukee would never have received such scrutiny. And if JetBlue's former flight attendant Steven Slater had popped that slide in Denver rather than at JFK, I sincerely doubt he would have become an instant media star.

Unfortunately, the same rules apply to safety. Some domestic airline incidents and accidents are more carefully scrutinized simply because of where they occur. Alaska Airlines underwent a virtual maintenance meltdown before and after the crash of Flight 261 in 2000, but outside the company's hometown of Seattle, this alarming story never gained traction on a national scale. Even though Alaska Airlines is ranked as a major carrier, its route map is noticeably thin in the Northeast, and therefore in my view it flies below a lot of media radar. And what's even more alarming is that a lack of press coverage can delay action on the part of the FAA, which has a reputation for responding to crises only after they're full-blown media events.

Flying Nonrevenue: Airline Wining and Dining

One Saturday in 1994 I took my son to the circus at Madison Square Garden. But we didn't squeeze into the bleachers with the hoi polloi; we sat high up in a private skybox, with kosher franks and blondies and other treats bought and paid for by KIWI International Air Lines. In its early stages, KIWI still had enough start-up funding left over to wine and dine the travel press, but that dried up fast enough.

For the aviation editor of a travel trade magazine, it was more than okay to accept such invitations. Actually, it was expected that I would spend my evenings and weekends being schmoozed by airline execs. In fact, it was a proud tradition: travel trade reporters had been receiving free lunches and tchotchkes since stagecoach operators decided to throw the first press trip.

Two years earlier I had left the airline industry to start at the bottom in journalism, and the pay cut was steep indeed. Some days I showed up at the magazine wondering if I had enough cash for lunch. A slice of pizza but walk home via the 59th Street Bridge? Inevitably I'd receive a fax that some airline or another was holding a press luncheon that day and within hours I'd be at Smith & Wollensky or Windows on the World or some other classic chophouse. With only four bucks in my wallet, I'd choose between House-Made Tagliatelle Formaggio or Porcini-Crusted Duck with Balsamic Reduction. Eventually I learned not to check a coat so I'd be spared the embarrassment of coming up short with a tip.

We trade reporters expected such treatment from airlines. So you can imagine my surprise when KIWI's private skybox door opened and in strolled the aviation editor for a major national publication, along with his wife and young daughter. His suddenly ashen face told the tale. There was no airline news being made at the circus, no exclusive interviews. This journalist was in flagrant violation of his publication's own stated policy on accepting freebies from corporations he wrote about. And clearly he had not expected a lowly trade reporter to be passing the blondies. I never squealed on him. But I learned early on that the divide was not nearly as great as I had imagined between trade journalists and those writing for the public.

And it's a good thing those tickets weren't charged to me, because there's no telling how big a tab I ran up on airline vouchers in the eight years I wrote for trades. I flew all over the world to airline press events, often in first class (business class when necessary). In Frankfurt, I stood on the tarmac as five Star Alliance wide-body commercial jets from five different airlines buzzed the field in formation. I flew on several inaugural flights, including the delivery of the first Boeing 777, thanks to United, and the first nonstop flight to Beijing, thanks to Northwest. In fact, airlines asked me to ride on every kind of commercial aircraft from helicopters to the Concorde.

Meanwhile, KLM invited me to tour the Van Gogh Museum in Amsterdam, Lufthansa invited me to experience Oktoberfest in Munich, and British Airways invited me to visit the London theater district. Northwest paid for me to hike to the top of the Great Wall in China. I took a Jeep tour of the Arizona desert sponsored by America West and Jet-Skied in Hawaii courtesy of TWA. I sailed the Hudson River on a party boat chartered by Continental and toured a Costa Rican coffee plantation thanks to Delta. Airline expense accounts allowed me to go clubbing in Lima, wine tasting in London's Soho, and snorkeling in Anguilla.

I had breakfast on El Al, lunch on Aer Lingus, and dinner on Alitalia. I consumed slaughterhouses of beef, barnyards of chicken breasts, boatloads of salmon, and barrels of red wine. And it was all charged to Singapore Airlines, Air Jamaica, Mexicana, Emirates, All Nippon Airways, SAS, Varig, Air Canada, Swissair, Thai Airways, and dozens of others. Although there was a long tradition of travel trade reporters calling travel agents "shrimp-eaters" who couldn't wait to attack an airline-paid buffet, quite a few crustaceans passed our own lips as well. Once—on a weeklong tour hosted by Airbus of the new A-340 factory in Toulouse, France—I and about a dozen other aviation journalists were fed enough foie gras to have killed Elvis.

The improprieties extended even further. One Friday evening at the Plaza Hotel in New York, I danced with an airline public relations rep at a company function and much later that night found myself making out with her at the Oak Bar. In my feeble defense, there were two mitigating factors: she worked for an outside PR agency, not the airline itself, and it was a small carrier based in a small country and rarely if ever would I be called upon to write about its operations. But then again, I knew more than one travel trade reporter who slept with airline employees they had interviewed.

Over the years I became friendly with many airline PR folks, and in a few cases I would even say I became friends. When I was struggling through a very painful custody battle, I often spoke for hours with a public relations rep at one of the major airlines, because she had endured a similar experience years before. She was a very kind person, and I don't believe for a moment she had ulterior motives for speaking to me; rather, the fact remains I was the one out of line, and turning to her for emotional support was wrong, just like the Oak Bar was wrong.

I share all this not as a mea culpa or a lame effort to cleanse my journalistic soul. I share it because in those years I think I was a pretty fair journalist and I broke quite a few important stories for the travel trade press, many of which the airlines were quite unhappy about. Yet I never once considered that I might have been biased in any way by the wining and dining I received. Sometimes my trade editors massaged, tweaked, rewrote, and killed my work, depending on whether I had offended the sensibilities of a given airline, aka advertiser. But I never considered myself complicit, or that I somehow allowed relations with airline personnel to influence what I wrote. In fact, if that had been suggested I would have been shocked, angered, and offended.

That said, a fair amount of industry cheerleading is not unusual from the airline press; even a respected publication such as *Aviation Week & Space Technology* can appear to have blinders firmly in place. In naming Continental CEO Jeffery Smisek its 2010 Person of the Year, the magazine referred to him as "appearing to distance his airline from the accident" of Colgan Air Flight 3407 in Buffalo and called this only a "hiccup." Smisek's congressional testimony enraged family members of the fifty people killed in that crash, and it's doubtful any of those next of kin would consider his actions a hiccup.

At times, beat reporters can tend to sound not like impartial chroniclers but like industry flacks. I asked Joe Brancatelli if journalists are susceptible to Stockholm Syndrome if they cover one business too long, and he said, "I think they are." He claimed the relationship can become incestuous, particularly when reporters are worried about access to key executives for one-on-one interviews.

However, an old friend who covered the airline beat from

Washington for many years says this of the "free lunch" issue: "I think there was a backlash against that kind of coziness after deregulation. If you're a good solid reporter who is fair, I don't think you will be able to be bought—especially that cheaply. I've always trusted that my friends [in the industry] understand that I have a job to do. Look, some of my best stories have come from those relationships." This reporter adds, "I understand there's a danger in getting too cozy. But with airlines and finance, you have to make an exception. It's too complex for new people to cover."

Entering the Nonprofit Monastery

The way I learned to curtail my relations with the airlines was to find religion and take vows. No, not that kind. I began writing for the largest nonprofit consumer organization in the world. In early 2000 I became the editor in chief of *Consumer Reports Travel Letter*, and found Consumers Union is unlike any other publisher in America. The company anonymously buys and tests all kinds of products, with no advertising, no marketing deals, no "strategic partnerships." Even the fund-raising department won't accept gifts from corporations. When *Consumer Reports* screws up, it's never because someone kissed up to an advertiser.

This is no small matter. The wall between editorial and sales is quite low at some media companies, and airline executives have been happy to vault it. As you read earlier, Delta's racist and homophobic interviewing practices in 1991 were highlighted by *Newsday* in the front-page article "They Love to Pry and It Shows." *American Journalism Review* later reported that the inflammatory headline prompted Delta to pull its ads from *Newsday*, costing the newspaper $2.2 million—a graphic case of an airline attempting to influence media coverage.

Fast-forward twenty years and it appears an entirely new management team at Delta still responds to negative coverage by shooting the messenger—with a cannon. In March 2011, Atlanta radio producer Chadd Scott encountered a lengthy delay on a Delta departure out of St. Louis so he tweeted: "The bean counter who saved Delta a few bucks in st. lou hoping he wouldn't need more de-icing fluid this year screwed a lot of people today." By the time Scott returned to Atlanta, he was fired; according to WXIA-TV, "his bosses told him Delta threatened to pull its advertising from the station."

Journalism ethics expert Fred Brown notes, "Is there pressure from advertisers? At good publications it shouldn't matter. The proper response is to resist that pressure." And such pressure can be positive as well as negative.

Back when I was interviewing at Consumers Union that holiday season I stopped in to see *CRTL*'s outgoing editor, Laurie Berger. On her desk was a large box of assorted gourmet brownies, courtesy of America West Airlines, which I recognized because that same morning the staff at my trade magazine had torn into an identical box. She explained they would overnight the entire box back to the airline's headquarters in Phoenix. I was impressed. But I also let her know the white chocolate raspberry brownie had been excellent.

I won't lie. I went through a little bit of a withdrawal process when I first started working for a nonprofit. No lunches. No cocktail parties. No trips. No tchotchkes. No schmoozing. It never quite reached the night-sweats phase. *Martinis . . . must have appletinis . . . Balthazar . . . must do prix-fixe lunch at Balthazar!* But many an afternoon I'd feel itchy and isolated and cut off from the travel industry shindigs occurring a few miles to the south in Manhattan. Gradually I learned I didn't need to schmooze with travel execs in order to write about their industry.

CrankyFlier.com's Snyder has taken an extra step. His site includes a Code of Ethics, which is clearly and easily explained: "My goal is to be an open book here." In addition to stating his personal policies, Snyder goes further and provides complete details on every offer extended to him by airlines and other travel companies. It makes for fascinating reading, and he explains why he does it: "Nothing annoys me more than being called a shill for the airlines. I know I have strong ethics, so why not share it with the world?" He adds that ethics are particularly critical for new media outlets, noting, "Anyone can start a blog and you have no way of knowing who they are. Newspapers have tended to be trustworthy, but with blogs—who knows?"

Unfortunately, not every media organization sets clear policies. Many companies willingly accept freebies from the travel industry. Others establish weak procedures that are not enforced. And some establish strict policies that are occasionally ignored.

Veteran journalist Fred Brown currently serves as vice chair of the Ethics Committee for the Society of Professional Journalists. I asked him about the particular hazards of reporting on the airline industry, since the temptations seem riper than on other beats. Brown responded, "This issue is kind of difficult for newspapers and websites to address because they use a lot of freelancers and freelancers don't have a budget to travel. You read about a new airplane and you wonder, is it an [Associated Press] story or a freelance story?" But he adds, "Ideally a credible publication or credible website should pay its own way."

Getting Social: New Media

If traditional media outlets have not always provided full disclosure on airline issues, thankfully other channels have emerged. Dave Carroll, the author of the YouTube mega-hit "United

Breaks Guitars," makes an important point: "The beauty of my situation is that social media has leveled the playing field—you can now be heard."

Indeed, just as in most arenas, democratic and technological forces have combined to bypass mainstream media outlets. Consider that recent breaking news reports of airline disasters have led with cell phone images and text messages from rescuers and survivors alike, notably when US Airways Flight 1549 ditched in the Hudson River adjacent to the world's media capital. It's a phenomenon seen with extended tarmac delays, too, as passengers trapped on aircraft have called local media to provide real-time updates from on board, thus elevating such events into even larger news stories. Meanwhile, no gripe is too arcane for the leveling influence of the Internet; back in 2007, the website The Consumerist even provided graphic photographic and editorial details of a Continental flight in which lavatory waste backed up and began seeping down the aisles of the airplane.

Kate Hanni, the passenger rights advocate who knows a thing or two about generating media coverage, also knows what she's up against with the airlines: "They use a lot of money to buy a lot of ink for what they want." But she maintains that passengers capitalizing on social media by photographing and tweeting about service problems help keep pressure on the industry. And then there are TripAdvisor, IgoUgo, and dozens of other travel advice sites that solicit and post tens of thousands of comments from passengers. Some of these amateur reviews are biased, some are unfair, and some are downright wacky. But few would deny they have the power to influence an airline booking decision.

The flip side is that new problems can be brought to light through new media. In asserting that coverage drives policy, safety expert Todd Curtis stresses this includes coverage from both "formal and informal networks." In fact, his AirSafe.com site includes a discussion about a photo taken in-flight by a Continental passenger that clearly shows loose and missing screws on the engine strut of a Boeing 737. As Curtis notes, "What you have now is a pervasive reach of media. Any [event] that happens is discussed. It's hard to hide a problem."

Epilogue: Fighting the Clock

THIS BOOK HAS BEEN A journey. I've now spent twenty-seven years in and around aviation, but through research and writing I learned an awful lot I never knew I didn't know. Ostensibly this book is about the airlines, but in fact it's about much more, as we struggle to define what America is becoming in 2012.

When Pan Am was in its death throes, one executive noted, "If we went into the funeral business, people would stop dying." I thought of those words as I visited the Southern California Aviation facility in Victorville, otherwise known as the aircraft boneyard. This Mojave Desert cemetery contains hundreds of discarded commercial airplanes, some still painted in the colors of Delta or United, some partially obscured, others completely whitewashed. Domestic flights are fuller than at any time since World War II, yet this dysfunctional industry can't find a way to improve customer service, put more skilled Americans to work, and keep planes flying.

The question emerges: Is it too late to save America's airline industry? I certainly hope not. I think it's worth noting that I have immersed myself in the business for more than a quarter of a century and I was stunned to uncover some of the findings shared in this book. Over a very short period of time, the airlines have radically changed how they do business. Now it's up to all of us to reverse some of those dangerous trend lines.

Two years ago, in early 2010, I began the outline of what would eventually become these pages, though I put the project on hold once I joined the FAAC. I was also traveling about three times a week from Connecticut to New Jersey to visit my mother, who was succumbing to the cruelty of Parkinson's disease and had begun reacting with suspicion and paranoia when any of her many children paid daily visits to her facility. One night in March she spotted me and recoiled, while I told her repeatedly not to worry, that it was Bill. Finally, I invoked the sobriquet I had been running from since 1973: "Mom . . . it's Billy."

Then I tried to repay a forty-five-year-old debt by settling in at her bedside to soothe her into sleep. Conversation lulled, so I started a new thread. "I've begun writing a book," I told her. "A man wants to help me sell it."

One eyebrow rose. "A book? About what?"

I leaned in close. "It's about airplanes."

She would be dead within weeks, just shy of her ninetieth birthday, but at that moment her features softened and it was apparent she was no more than six going on seven. She had beaten diphtheria and was perched atop her father's shoulders, anxiously awaiting a glimpse of Lucky Lindy himself. America had oceans to conquer. "Bill?"

I smiled. "Yep. I was named for your dad."

She seemed surprisingly lucid then, and she nodded. "Make it a good book."

I kissed her on the forehead and dimmed the headboard light. "I'll try," I promised.

Conclusion: A Manifesto for Taking Back Our Skies

IN SEPTEMBER 2011 THE MEMBERS of the Future of Aviation Advisory Committee were called to Washington by Secretary LaHood. There we discovered that the collection of proposals we submitted the previous December was very much a work in progress. To the DOT's credit, there were advancements on passenger rights issues, particularly new rules to enforce transparency of airline pricing and added fees. And a forum was convened to spur a long-term discussion between airline executives and labor leaders. But many of what I view as the most critical safety issues—particularly enhanced oversight of outsourced maintenance and regional carriers—never made it past the discussion stage. As for child restraint systems, the FAA backed away from mandating CRSs on domestic airlines, and instead said it will continue bolstering efforts to educate parents.

I've come to see that the deepest problems facing the U.S. airline industry are the deepest problems facing the United States itself. These include the pervasive effects of corporate influence throughout all three branches of government; political gridlock; the ever-widening economic gap between corporate CEOs and rank-and-file workers; federal regulators failing to regulate the industries they are sworn to oversee; the wholesale outsourcing of decent jobs, particularly outside the United

States; and a willful ignorance of how industry is affecting the environment.

Quite a few of the experts I interviewed for this book offered sensible, practical, and reasonable solutions to address the many problems facing the U.S. airline industry. After analyzing and weighing a broad spectrum of competing philosophies, I came to form my own proposals for fixing the airlines. Therefore, I respectfully suggest the following actions be considered.

• Crafting a National Transportation Policy

A hot topic on the Future of Aviation Advisory Committee was "intermodalism," a term meant to encompass how various transportation modes intersect. But a committee composed solely of aviation experts charged with examining aviation is in itself lacking—how can those with a vested interest in promoting the airline industry help determine the big-picture transportation needs of the United States? The airline lobby has its cheerleaders, as do the lobbies for highways, rail, motor coaches, etc. Consider that one expert I spoke to noted that for many U.S. airports, the largest single source of revenue is parking. So what incentive does aviation have to promote Amtrak or intercity buses? If the U.S. Department of Transportation truly wants to examine the future of aviation, then the White House must instruct the DOT to examine the future of all transportation modes at once, so we as a nation can discuss transportation not only in terms of convenience, but also in the context of jobs, the environment, infrastructure, and the public good. As Kevin Mitchell has noted, it's past time for America to develop a true national transportation policy.

• It's Time for Partial Reregulation

The Airline Deregulation Act of 1978 held out many promises, and the industry would have us believe those promises have been fulfilled, primarily because of the tremendous influx in commercial flying over the last thirty-three years. Prior to 1978, airline safety was improving, more passengers were flying, and fares were decreasing. What remains unknown and unknowable is how many of the changes we have seen since 1978 would have occurred naturally, even in a regulated environment, due to technological advancements and other big-picture factors. Today it's fair to ask, who exactly has benefited from airline deregulation? Investors have lost billions over three decades. Airline employees have been downsized, outsourced, offshored, and cut back. Communities have lost service and in some cases airlines. And as for consumers, they are more frustrated with commercial aviation than they have ever been. The claim that base airfares have not risen dramatically in the aggregate conceals the billions now paid by passengers in ancillary fees and does not address the inequities in a system that primarily has seen fares repressed only by the presence of low-cost carriers. The industry's stellar safety record is threatened by cost cutting and outsourcing. It's time to acknowledge the simple fact that the airlines are run as a cutthroat free-market exercise, yet they provide a necessary and vital service for the nation and are intrinsic to its economy, welfare, security, and even defense. Airline executives make individual decisions that may seem logical on a corporate basis but are detrimental to the country on a macro level. As former American Airlines chairman Bob Crandall notes, the airline industry is a necessary utility, and as such it requires the regulation of a utility. Few would seek a complete return to 1978, when the Civil Aeronautics Board micromanaged the industry, route by route and fare by fare. But it is time for the White House and Congress to direct the Department of Transportation to reconsider how partial reregulation will level the industry's inherent deficiencies. This means ensuring that communities and passengers are served, fares are priced to ensure a return on investment, labor agreements are fair to both parties, and safety is not compromised.

• Justice Department Should Enforce Fair Competition

The architects of the Airline Deregulation Act made it clear that success would be dependent on a robust U.S. Department of Justice that would enforce fair competition and sanction predatory pricing and other forms of illegal anticompetitive behavior and prevent rapid industry consolidation that would also threaten competition and long-standing antitrust law. In recent years the Justice Department has failed on both counts, by allowing predatory practices that have driven some low-fare carriers out of existence and by allowing mergers and acquisitions that have created a "too big to fail" scenario for the domestic airline industry. It's time the White House directed the department to fulfill its mandate to American consumers.

• Meeting International Standards for Emissions

The U.S. aviation industry—and at times the U.S. government—has fought international efforts to codify and address commercial aviation's carbon emissions and environmental footprint. The White House and Congress need to work with rather than oppose other governments seeking to incentivize greener behavior on the part of the airline industry.

• NTSB Should Have Teeth

It's shameful that recommendations on aviation safety put forth by the National Transportation Safety Board to the Federal Aviation Administration can remain on the NTSB's "Most Wanted" list for years and even decades. If the FAA fails to act on such recommendations, then the NTSB should be empowered to forward such proposals directly to Congress for immediate consideration.

• Aircraft Maintenance Should Be Performed Domestically

The airline industry's dramatic experimentation with maintenance outsourcing has created a dangerous environment in which frontline FAA inspectors claim they do not have the resources to properly oversee the work being done. All aircraft maintenance for U.S. airlines should be performed in the United States, and the White House and Congress should ensure that the FAA has the budget, staffing, and resources to properly oversee this work.

• A Higher Standard for Regional Airlines

Although the Federal Aviation Regulations stipulate a single standard for mainline and regional air carriers, the regulations address a minimum proficiency that has proven to be lacking in some cases. In practice, many mainline airlines maintain higher standards for hiring, training, and operations than their regional marketing partners. The FAA should examine the federal standards to determine if the minimums should be raised.

• More Transparency for Foreign Airlines

Critics claim the U.S. State Department has more authority than the DOT in determining which foreign air carriers operate into the United States. The European Union provides consumers with full transparency by identifying every banned airline by name; the United States should adopt a similar model.

• Fund NextGen

Political gridlock led to the shameful state of twenty-one consecutive failures by Congress to provide long-term funding for the FAA, which in turn has stalled the implementation of the Next Generation Air Transportation System, the desperately needed revamp of the nation's outdated air traffic control network. NextGen needs funding—now.

- **Revamp Chapter 11 Bankruptcy Laws**

Far from viewing the process as a stigma, in recent years airline executives have embraced bankruptcy reorganization on a broad scale. In the end, Chapter 11 can save companies, jobs, and communities. But some airline executives have used bankruptcy as a means to secure sacrifices from labor and other stakeholders—while boosting their own economic portfolios. Bankruptcy laws should be amended so that the senior executives who oversaw the entry into Chapter 11 are replaced, and senior management is not excessively rewarded while other workgroups sacrifice.

- **A Facility for Passengers Banding Together**

Unfortunately, despite their overwhelming numbers, consumers have less of a voice in setting domestic aviation policy than any other group of stakeholders. Federal preemption effectively removes many of the basic rights customers are afforded throughout American commerce. Ralph Nader has long advocated for an effective organization that would allow consumers to be heard and would facilitate their grievances.

- **A Codified Passenger Bill of Rights**

The airlines' Contracts of Carriage have become ineffectual, obtuse, and impossible for consumers to comprehend, let alone enforce. The DOT needs to instate a new Rule 240 for the postderegulation environment: A transparent and uniform set of standards for what carriers need to provide in the case of extended flight delays, flight cancellations, and involuntary denied boardings. The European Union has done this for its passengers, and the resulting rules are clear, cogent, and easy for all passengers to understand.

- **Address Airport Slots as Public Property**

As Stephen Van Beek notes, private companies such as airlines and banks buy and sell public property that belongs to all Americans. Airport takeoff and landing slots should not be treated as corporate assets, and in fact the DOT should reexamine its process for awarding such slots. This means rewarding the public's greater good—airlines that provide more seats and therefore reduce carbon emissions.

Will perfect regulation create a perfect airline market? Sadly, no, because both are impossible to attain. But well-meaning and carefully considered imperfect regulation can help create a much more perfect airline market than we have today. The time has come to reclaim our skies.

Appendix A: Domestic Mainline Airline Partnerships with Regionals

(AS OF DATE OF PUBLICATION)

- Alaska Airlines partners with Horizon Air.
- American Airlines partners with American Eagle as well as an AmericanConnection carrier, Chautauqua Airlines.
- Continental Airlines maintains two sets of partnerships: Continental Connection flights are operated by Cape Air, Colgan Air, CommutAir, and Gulfstream International Airlines, while Continental Express flights are operated by Chautauqua Airlines and ExpressJet Airlines.
- Delta Air Lines partners with eight regionals: Atlantic Southeast Airlines, Chautauqua Airlines, Comair, Compass, Mesaba Airlines, Pinnacle Airlines, Shuttle America, and SkyWest Airlines.
- Frontier Airlines operates Frontier Express flights with partners Chautauqua Airlines and Republic Airlines.
- United Airlines has eight United Express regional partners: Atlantic Southeast Airlines, Colgan Air, ExpressJet, GoJet Airlines, Mesa Airlines, Shuttle America, SkyWest Airlines, and Trans States Airlines.
- US Airways Express consists of nine regional carriers: Air Wisconsin, Chautauqua Airlines, Colgan Air, Mesa Airlines, Mesaba Airlines, Piedmont Airlines, PSA Airlines, Republic Airlines, and Trans States Airlines.

Appendix B: Domestic Regional Airline Partnerships with Mainlines

(AS OF DATE OF PUBLICATION)

- Atlantic Southeast partners with both Delta and United.
- Chautauqua partners with five mainlines: American, Continental, Delta, US Airways, and sister company Frontier.
- Colgan partners with Continental, United, and US Airways.
- ExpressJet partners with both Continental and United.
- Mesa partners with both United and US Airways.
- Mesaba partners with both Delta and United.
- Republic Airlines partners with both sister company Frontier and US Airways.
- Shuttle America partners with both Delta and United.
- SkyWest partners with both Delta and United.
- Trans States partners with both United and US Airways.

Glossary

A4A: Airlines for America, formerly known as the Air Transport Association of America, the nation's largest airline trade organization

CAB: Civil Aeronautics Board, a former government agency that was charged with managing the nation's airlines and was shut down in 1984

DHS: U.S. Department of Homeland Security, a cabinet-level department responsible for securing the nation from terrorist threats

DOJ: U.S. Department of Justice, a cabinet-level department charged with overseeing antitrust and competition issues in U.S. industries

DOT: U.S. Department of Transportation, a cabinet-level department charged with overseeing aviation and other transportation modes

FAA: Federal Aviation Administration, a division of the DOT specifically charged with overseeing commercial and civil aviation safety and operating the nation's civil air traffic control network

FAAC: Future of Aviation Advisory Committee, a committee formed by the DOT in 2010 to examine the state of the U.S. airline industry, on which the author served

IATA: International Air Transport Association, the world's largest airline trade organization

ICAO: International Civil Aviation Organization, a United Nations–chartered organization dedicated to the "safe and orderly growth" of civil aviation

NTSB: National Transportation Safety Board, an independent government agency charged with investigating transportation safety and accidents

TSA: Transportation Security Administration, a division of the DHS responsible for oversight of security measures within the nation's transportation infrastructure

Selected Bibliography

Books

Breyer, Stephen. *Regulation and Its Reform.* Cambridge, MA: Harvard University Press, 1982.

Crouch, Tom. *The Bishop's Boys: A Life of Wilbur and Orville Wright.* New York: W.W. Norton and Company, 1989.

Curtis, Todd. *Understanding Aviation Safety Data: Using the Internet and Other Sources to Analyze Air Travel Risk.* Warrendale, PA: SAE International, 2000.

Dempsey, Paul Stephen, and Andrew R. Goetz. *Airline Deregulation and Laissez-Faire Mythology.* Westport, CT: Quorum Books, 1992.

Kahn, Alfred E. *The Economics of Regulation: Principles and Institutions.* New York: John Wiley and Sons, 1971.

Mueller, Scott T. *The Empty Carousel: A Consumer's Guide to Checked and Carry-on Luggage.* Longwood, FL: Millkot Publishing, 2008.

Nader, Ralph, and Wesley J. Smith. *Collision Course: The Truth About Airline Safety.* Blue Ridge Summit, PA: McGraw-Hill/TAB Books, 1994.

Ott, James, and Raymond E. Neidl. *Airline Odyssey: The Airline Industry's Turbulent Flight into the Future.* New York: McGraw-Hill, 1995.

Peterson, Barbara Sturken, and James Glab. *Rapid Descent: Deregulation and the Shakeout in the Airlines.* New York: Simon and Schuster, 1994.

Petzinger, Thomas Jr. *Hard Landing: The Epic Contest for Power and Profits That Plunged the Airlines into Chaos.* New York: Times Books, 1995.

Schiavo, Mary. *Flying Blind, Flying Safe.* New York: Avon Books, 1997.

Sullenberger, Capt. Chesley, and Jeffrey Zaslow. *Highest Duty: My Search for What Really Matters*. New York: HarperCollins, 2009.

Newspaper/Magazine/Online Articles

Adcock, Sylvia, "FAA Probes Near-Miss at MacArthur," *Newsday*, March 25, 2000.

Associated Press, "Alaska Airlines Facing FBI Probe," March 17, 2000.

———, "Rerouting Work to Mexico," February 28, 2004.

———, "FAA Let Northwest Airlines Flout Safety Orders," July 23, 2010.

———, "FAA Cites Improper Repairs to Southwest Planes," September 12, 2011.

Belden, Tom, "US Airways' Big Hiccup," Philly.com, March 5, 2007.

Blair, Jayson, "Fog Delay of 3 Hours Hits JetBlue in First Flight," *New York Times*, February 12, 2000.

Bly, Laura, "Child Restraint Seats on Airlines: What Price Safety?" United Press International, June 3, 1990.

Brancatelli, Joe, The Brancatelli File, 2005–2006 (numerous).

Breyer, Stephen. "Airline Deregulation, Revisited, " *Businessweek*, January 20, 2011.

CNN.com, "TSA Denies Having Required a 95-Year-Old Woman to Remove Diaper," June 27, 2011.

Compart, Andrew, "Executive Pay and U.S. Airlines," *Aviation Week and Space Technology*, February 6, 2009.

C-SPAN, "Carnegie Institute for Science: Bruce Schneier on Airport Security," January 6, 2011.

Daily Mail, "Florida Man Hauled Off Flight for 'Pleasuring Himself While in His Seat,'" May 24, 2011.

Eckert, Beverly, "My Silence Cannot Be Bought," *USA Today*, December 19, 2003.

Fortune magazine, "A CEO Perk Parade," June 24, 2011.

Gaynor, Tim (editor), "JetBlue Offers $4 L.A.-Area 'Carmageddon' Flights," Reuters, July 13, 2011.

Greenberg, Peter, "Navigating Airfares and Frequent Flier Programs," PeterGreenberg.com, July 21, 2011.

Harris, Byron, "Aircraft Repair Jobs Sold to Foreign Workers, Résumés Not Important," WFAA-TV, July 8, 2009.

Hawley, Chris, "Who Owns That Airplane? There's a Good Chance the FAA Has No Idea," Associated Press, December 10, 2010.

Huber, Mark, "How Things Work: Evacuation Slides," *Air and Space/Smithsonian*, October–November 2007.

Jackson, David S., John S. DeMott, and Allen Pusey, "Dirty Tricks in Dallas," *Time*, March 7, 1983.

Katovsky, Bill, "Flying the Deadly Skies: Whistleblower Thinks the State of U.S. Aviation Security Invites Another Attack," *San Francisco Chronicle*, July 9, 2006.

LaHood, Ray, "Tarmac Rule Worked as Planned," *Newark Star-Ledger*, March 14, 2011.

Langewiesche, William, "The Lessons of ValuJet 592," *Atlantic Monthly*, March 1998.

Levin, Alan, "Airlines Seek to Move Air Marshals from First Class," *USA Today*, October 19, 2010.

Ling, P., "Airline CEO Compensation Roundup," UpTake Networks, May 4, 2010.

Lowy, Joan, "Air Traffic Control Error Numbers Double," Associated Press, February 11, 2011.

Marks, Alexandra, "Air Marshals Stretched Thin," *Christian Science Monitor*, December 28, 2005.

Maxon, Terry, "Safety Board Finds Rivet Problems on Damaged Southwest Airlines Jet," *Dallas Morning News*, April 25, 2011.

Morrow, Lance, "Airline Pollution: The Sky Has Its Limits," *Time*, May 7, 2001.

Mutzabaugh, Ben, "Air France Stewardess Stole from Passengers While They Slept," USAToday.com, July 21, 2010.

———, "Spirit Airlines: Have You Seen Our Weiner?" USAToday.com, June 8, 2011.

Noonan, Peggy, "We Pay Them to Be Rude to Us," *Wall Street Journal*, August 13, 2010.

Public Broadcasting System/Frontline, "Flying Cheap," February 9, 2010.

———, "Flying Cheaper," January 18, 2011.

Peterson, Barbara S., "The Great Escape," *Condé Nast Traveler*, November 2005.

———, "Inside Job: My Life as an Airport Screener," *Condé Nast Traveler*, March 2007.

Phillips, Don, "Improper Use of Tape to Fix Wings May Lead to FAA Fine for United," *Washington Post*, December 4, 2002.

Pogash, Carol, "General Mills' Gift to Journalism," *American Journalism Review*, July/August 1995.

Popken, Ben, "An Interview with the Fee-Happy CEO of Spirit Airlines," *Consumerist*, July 14, 2011.

Ranson, Lori, "Congress Reviews Non-Certified Repair Stations," *Aviation Week and Space Technology*, March 30, 2007.

Schlangenstein, Mary, "United Air to Move 165 Call-Center Jobs Back to U.S.," Bloomberg News, February 10, 2009.

Shannon, Darren, "2010 Person of the Year: Jeffrey A. Smisek," *Aviation Week and Space Technology*, January 3, 2011.

Stoller, Gary, "Spirit Airlines Is Cheap, and CEO Ben Baldanza's Proud of It," *USA Today*, June 22, 2009.

———, "Planes with Maintenance Problems Have Flown Anyway," *USA Today*, February 4, 2010.

———, "How Safe Is That Foreign Airline?" *USA Today*, June 17, 2011.

Swelbar, William, "A Look at U.S. Airline CEO Compensation Through a Different Lens," Swelblog.com, February 11, 2009.

TheSmokingGun.com, "Smoking Woman in Air Rage," June 19, 2008.

Time, "Aviation: Trippe's Big Bid," December 28, 1962.

———, "New Era in the Air: Cheap Fares, Crowded Flights," August 14, 1978.

Wade, Betsy, "Flying Blind: An Unexpected Change of Airline," *New York Times*, October 21, 1990.

WSB-TV, "Delta Rolls Out Platinum Upgrades for Top Republicans," May 16, 2011.

WXIA-TV, "Atlanta Sports Radio Producer Fired for Tweets," March 16, 2011.

Research/Journals/White Papers/Testimony/Presentations/ Releases

Adams, Rob, "The Effects of Aviation Noise on People," Landrum and Brown, March 5, 2006.

Aeronautical Repair Station Association, "Industry Economic Data," undated.

AeroStrategy Management Consulting, "Global MRO Market Economic Assessment," August 21, 2009.

Air Line Pilots Association International, "Creating One Level of Safety for Both Passenger and Cargo Carriers," undated.

———, "Runway Incursions: A Call for Action," March 2007.

———, "Recommendations to Improve the Federal Flight Deck Officer Program," July 2007.

———, "Secondary Flight Deck Barriers and Flight Deck Access Procedures: A Call for Action," July 2007.

Air Transport Association, "U.S. Airline Bankruptcies and Service Cessations," undated.

———, "U.S. Airline Mergers and Acquisitions," undated.

———, "When America Flies, It Works: 2010 Economic Report," 2010.

———, "The Economic Climb-Out for U.S. Airlines: Global Competitiveness and Long-Term Viability," August 8, 2011.

Airbus, "World's First Scheduled Passenger Biofuel Flights Commence," July 15, 2011.

Alonso, Juan, Ian Waitz, et al, "Greening U.S. Aviation," Presentation at United Nations Climate Change Conference, December 2009.

AltAir Fuels, "AltAir Fuels Partners with USDA to Spur Camelina Growth," July 26, 2011.

American Civil Liberties Union, "Technology and Liberty: Airport Security," November 17, 2010.

American Customer Satisfaction Index, "Passenger Discontent Pervasive for Airlines," June 2011.

Anderson, Sarah, et al, "Executive Excess 2008: How Average Taxpayers Subsidize Runaway Pay," Institute for Policy Studies and United for a Fair Economy, August 25, 2008.

Bailey, Elizabeth E., "Air Transportation Deregulation," Wharton School, 2008.

Barnett, Arnold, "Cross National Differences in Aviation Safety Records," *Transportation Science*, 2010.

Boeing Commercial Airplanes, "Aviation and Sustainable Biofuel," 2011.

———, "Statistical Summary of Commercial Jet Airplane Accidents, Worldwide Operations, 1959–2010," June 2011.

Borenstein, Severin, "On the Persistent Financial Losses of U.S. Airlines: A Preliminary Exploration," National Bureau of Economic Research, 2011.

Bronzaft, Arline, "United States Aviation Transportation Policies Ignore the Hazards of Airport-Related Noise," World Transport Policy and Practice, 2003.

Buffett, Warren E., Berkshire Hathaway Inc., Letter to Shareholders, February 2008.

Business Travel Coalition, "Predatory Airline Behavior Is Back," March 17, 2011.

———, "A Departure from the Ordinary: Presentation for Department of Transportation," June 29, 2011.

———, "U.S. Trusted Traveler Program Doomed Unless Flaws Are Addressed," July 15, 2011.

Carlson Wagonlit Travel, "Travel Management Priorities for 2011: Insights into the Rebound," January 2011.

The Center for Public Integrity, "Broken Government: An Examination of Executive Branch Failures Since 2000," January 7, 2009.

Chapin, Robert, "Runway Incursions 2000–2010: Is Safety Improving?" Eastern Michigan University, 2010.

Commercial Aviation Safety Team, "Wrong Runway Departures," August 2007.

Cooper, Mark N., "Freeing Public Policy from the Deregulation Debate: The Airline Industry Comes of Age," Consumer Federation of America, January 22, 1999.

Crandall, Robert L., "The Customer Is the Focus of Everything," Remarks to IATA Commercial Strategy Symposium, Istanbul, 2010.

Dempsey, Paul Stephen, "The Financial Performance of the Airline Industry Post-Deregulation," *Houston Law Review*, Vol. 45, No. 2, 2008.

Dudley, Susan E., "Alfred Kahn, 1917–2010: Remembering the Father of Airline Deregulation," *Regulation*, Spring 2011.

European Commission, "List of Airlines Banned Within the E.U.," April 21, 2011.

European Federation for Transport and Environment, "Report Shows Fifty-Year Failure of Aviation Industry to Improve Fuel Efficiency," December 7, 2005.

Fischer, John W., Bart Elias, and Robert S. Kirk, "U.S. Airline Industry: Issues and Role of Congress," Federal Publications, April 2008.

Flightglobal.com, "Fleet Watch," June 2011.

Furton, Kenneth G., and Douglas P. Heller, "Advances in the Reliable Location of Forensic Specimens Through Research and Consensus: Best Practice Guidelines for Dog and Orthogonal Instrumental Detectors," *Canadian Journal of Police and Security Services*, June 2005.

Goodpaster, Kenneth E., and David E. Whiteside, "Braniff International: The Ethics of Bankruptcy," Harvard Business School, 1984.

Greenspan, Alan, "Remarks at the Ronald Reagan Library: The Reagan Legacy," April 9, 2003.

Hagan, Bill, "How to Survive a Plane Crash," AmSafe Aviation Blog, December 25, 2009.

Hassin, Orit, and Oz Shy, "Codesharing Agreements, Frequency of Flights, and Profits in the Airline Industry," University of Haifa, 2000.

Hazel, Bob, Aaron Taylor, and Andrew Watterson, "Airline Economic Analysis for the Raymond James Global Airline Conference," Oliver Wyman, 2011.

Henderson, M. Todd, "Paying CEOs in Bankruptcy: Executive Compensation When Agency Costs Are Low," *Northwestern University Law Review*, Vol. 101, No. 4, 2007.

Holey, Tim, and Kendall Krieg, "Strategic Approach to Fire Safety," Boeing Commercial Airplanes, June 2005.

Horan, Hubert, "The Anti-Competitive Risks of a Delta-Northwest Merger and Extreme Consolidation of Intercontinental Airlines," U.S. House Aviation Subcommittee, 2008.

Hoyland, Tim, Chris Spafford, and Roger Lehman, "MRO Industry Landscape 2010: A New Level of Competition Emerges," Oliver Wyman, 2010.

Hu, Xing, René Caldentey, and Gustavo Vulcano, "Revenue Sharing in Airline Alliances," New York University, November 2, 2010.

Hudson, Paul S., "Airline Passenger Tarmac Confinements and Delays: Reasonable Regulation Trumps Market Forces," *Air and Space Lawyer,* American Bar Association, 2010.

IdeaWorks, "Five Frequent Flyer Programs Accrue Revenue of $7 Billion," September 12, 2011.

IdeaWorks and Amadeus, "Review of Ancillary Revenue Results," May 31, 2011.

Insurance Institute of America, "Odds of Death in the United States by Selected Cause of Injury," 2007.

International Air Transport Association, "Financial Impact of Extending the EU Emissions Trading Scheme to Airlines," January 9, 2007.
———, "Aircraft Accident Rate Is Lowest in History—Still Room for Improvement, Regional Concerns Remain," February 23, 2011.
International Civil Aviation Organization/Environment Branch, "Aircraft Engine Emissions," undated.
———, "Aircraft Noise," undated.
Kasper, Daniel M., and Darin Lee, "Why Airline Antitrust Immunity Benefits Consumers," *GCP: The Antitrust Chronicle*, September 2009.
Khakh, M. Usman, "Alaska Airlines and Flight 261," Harvard Business School, March 26, 2001.
Lardner, James, and Robert Kuttner, "Flying Blind: Airline Deregulation Reconsidered," Demos, 2009.
Levine, Michael E., "Regulation, the Market, and Interest Group Cohesion: Why Airlines Were Not Reregulated," New York University School of Law, 2006.
Lipinski, Rep. Dan, "Lipinski Slams FAA for Lapses in Airline Safety Inspections," U.S. House of Representatives, April 3, 2008.
Mann, Robert, "One Industry, One Standard of Safety," Coalition of Airline Pilots Associations, 2009.
———, "Uncertain Times for an Uneasy Industry," American Bar Association Forum on Air and Space Law, 2009.
McKenzie, Dan, "Global Market Share Handbook," Hudson Securities, February 15, 2011.
———, "Airline Industry Changes and the Corporate Travel Manager," Hudson Securities, May 19, 2011.
Moss, Diana L., "Airline Mergers at a Crossroads: Southwest Airlines and AirTran Airways," American Antitrust Institute, December 14, 2010.
National Air Traffic Controllers Association, "NATCA: A History of Air Traffic Control," undated.
OpenSecrets.org, "Lobbying—Top Industries," 2011.

Oster, Clinton V., Jr., and John S. Strong, "Predatory Practices in the U.S. Airline Industry," Indiana University and College of William and Mary, 2001.

PlaneStats.com, "Global Capacity Trends by Region and Equipment Type," 2010.

PricewaterhouseCoopers/India, "Measuring Human Capital—Driving Business Results," March 27, 2011.

Project on Government Oversight, "Breaking the Sound Barrier: Experiences of Air Marshals Confirm Need for Reform at the OSC," November 25, 2008.

Public Citizen, "Delay, Dilute, and Discard: How the Airline Industry and the FAA Have Stymied Aviation Security Recommendations," October 2001.

Quigley, Claire, et al., "Anthropometric Study to Update Minimum Aircraft Seating Standards," ICE Ergonomics, et al., July 2001.

Regional Airline Association, "FAA Forecasts Regional Carrier Enplanements to Reach 295.9 Million by 2031," February 16, 2011.

Shane, Jeffrey N., and Warren L. Dean Jr., "Alliances, Immunity, and the Future of Aviation," *Air and Space Lawyer*, American Bar Association, 2010.

SITA, "Baggage Report 2011," March 30, 2011.

Spafford, Chris, Roger Lehman, and Tim Hoyland, "Aviation MRO: The Next Place to Land for Private Equity Investors," Oliver Wyman, 2008.

Spafford, Chris, Roger Lehman, and Brian Prentice, "In Airframe Maintenance, Building Real Partnerships Will Benefit Both Airlines and Suppliers," Oliver Wyman, 2009.

Townsend, Corky, Boeing Commercial Airplanes, "Enhancing Crashworthiness," 2011.

Transportation Research Board, "Entry and Competition in the U.S. Airline Industry: Issues and Opportunities," 1999.

Vincent, Billie H., "Have Airports, Airlines, and Governments Done Enough (or Too Much) to Avoid Another Catastrophic Terrorist

Event?" McGill University Worldwide Conference on Current Challenges in International Aviation, September 25, 2004.

World Health Organization, "Guidelines for Community Noise," 1999.

U.S. Government Sources

National Transportation Safety Board, Accident Reports 1967–2011 (numerous).

———, "Emergency Evacuation of Commercial Airplanes," June 27, 2000.

———, "Safety Recommendation: Revoked Repair Stations," February 9, 2004.

———, "Safety Recommendation: Child Restraint Systems," August 11, 2010.

U.S. Department of Homeland Security/Office of Inspector General, "Statement of J. Richard Berman Before the Select Committee on Homeland Security/ U.S. House of Representatives," October 8, 2003.

———, "Audit of Passenger Processing Reengineering," March 2004.

———, "Statement of Clark Kent Ervin Before the Committee on Transportation and Infrastructure/Subcommittee on Aviation/ U.S. House of Representatives," April 22, 2004.

———, "Audit of Passenger and Baggage Screening Procedures at Domestic Airports," September 2004.

———, "Review of the TSA Passenger and Baggage Screening Pilot Program," September 2004.

———, "Follow-Up Audit of Passenger and Baggage Screening Procedures at Domestic Airports (Unclassified Summary)," March 2005.

———, "A Review of Procedures to Prevent Passenger Baggage Thefts," March 2005.

———,"Transportation Security Administration's Procedures for Law Enforcement Officers Carrying Weapons Onboard Commercial Aircraft (Unclassified Summary)," September 2005.

———, "Transportation Security Administration's Revised Security Procedures (Unclassified Summary)," September 2005.

———,"Review of the Transportation Security Administration's Use of Pat-Downs in Screening Procedures," November 2005.

———,"Review of the Transportation Security Administration's Management Controls Over the Screener Recruitment Program," December 2005.

———,"Review of the TSA Non-Screener Administrative Positions," September 2006.

———,"Review of Allegations Regarding San Francisco International Airport," October 2006.

———,"Improvements Needed in TSA's Federal Flight Deck Officer Program," December 2006.

———,"Audit of Access to Airport Secured Areas (Unclassified Summary)," March 2007.

———,"Transportation Security Administration's Oversight of Passenger Aircraft Cargo Security Faces Significant Challenges (Redacted)," July 2007.

———,"A Follow-Up Review of the Transportation Security Officer Background Check Process," August 2007.

———,"Transportation Security Administration Vetting of Airmen Certificates and General Aviation Airport Access and Security Procedures," July 2011.

———,"TSA's Oversight of Airport Badging Process Needs Improvement (Redacted)," July 2011.

U.S. Department of Homeland Security/Transportation Security Administration, "TSA Makes Deep Cuts in Backlog of Checked Baggage Claims from Passengers," September 20, 2004.

———, "Review of Alleged Actions by Transportation Security Administration to Discipline Federal Air Marshals for Talking to the Press, Congress, or the Public," November 2004.

———, "Review of Department's Handling of Suspicious Passengers Aboard Northwest Flight 327 (Redacted)," March 2006.

U.S. Department of Transportation/Aviation Consumer Protection Division, Monthly Air Travel Consumer Reports 1987–2011 (numerous).

———, "Air Travelers: Tell It to the Judge/A Consumer's Guide to Small Claims Courts," September 1994.

U.S. Department of Transportation/Federal Aviation Administration, "Speech: The Spirit of December 14th," February 20, 2003.

———, "Annual Runway Safety Report 2010," 2010.

———, "Safety Considerations When Flying with Children," December 9, 2010.

———, "Use of Child Restraint Systems on Aircraft," December 9, 2010.

———, "Instrument Procedures Handbook," March 9, 2011.

U.S. Department of Transportation/Federal Aviation Administration/Air Traffic Organization, "Aviation Noise Effects," March 1985.

———, "The Economic Impact of Civil Aviation on the U.S. Economy," December 2009.

U.S. Department of Transportation/Future of Aviation Advisory Committee, "Final Report," April 11, 2011.

U.S. Department of Transportation/Independent Review Team, Report of a Blue Ribbon Panel Appointed by Secretary Mary E. Peters: "Managing Risks in Civil Aviation: A Review of the FAA's Approach to Safety," September 2, 2008.

U.S. Department of Transportation/Office of Aviation Analysis, "The Southwest Effect," May 1993.

U.S. Department of Transportation/Office of Inspector General, "Report on Travel Agent Commission Overrides," March 2, 1999.

———,"FAA Oversight of Passenger Aircraft Maintenance," April 11, 2002.

———, "Review of Air Carriers' Use of Aircraft Repair Stations," July 8, 2003.

———, "Safety Oversight of an Air Carrier Industry in Transition," June 3, 2005.

———, "Air Carriers' Use of Non-Certificated Repair Facilities," December 15, 2005.

———, "Observations on FAA's Oversight of Aviation Safety," September 20, 2006.

———, "Aviation Safety: FAA's Oversight of Outsourced Maintenance Facilities," March 29, 2007.

———, "Key Safety and Modernization Challenges Facing the Federal Aviation Administration," April 17, 2008.

———, "Air Carriers' Outsourcing of Aircraft Maintenance," September 30, 2008.

U.S. General Accounting Office "Aircraft Maintenance: Additional FAA Oversight Needed of Aging Aircraft Repairs," May 1991.

———, "Commuter Airline Safety Would Be Enhanced with Better FAA Oversight," March 17, 1992.

———, "Effects on Changes in How Airline Tickets Are Sold," July 1999.

———, "Impact of Changes in the Airline Ticket Distribution Industry," July 2003.

———, "Federal Air Marshal Service Is Addressing Challenges of Its Expanded Mission and Workforce, but Additional Actions Needed," November 2003.

———, "Aviation Safety: Better Management Controls Are Needed to Improve FAA's Safety Enforcement and Compliance Efforts," July 2004.

U.S. Government Accountability Office, "Clear Policies and Oversight Needed for Designation of Sensitive Security Information," June 2005.

———, "Aviation Safety: Oversight of Foreign Codeshare Safety Program Should Be Strengthened," August 5, 2005.

———, "Aviation Safety: System Safety Approach Needs Further Integration into FAA's Oversight of Airlines," September 2005.

———, "Federal Air Marshal Service Could Benefit from Improved Planning and Controls," November 2005.

———, "Aviation Safety: FAA's Safety Efforts Generally Strong but Face Challenges," September 20, 2006.

———, "Progress Made in Systematic Planning to Guide Key Investment Decisions, but More Work Remains," February 13, 2007.

———, "Progress Report on Implementation of Mission and Management Functions," August 2007.

———, "Efforts to Strengthen Aviation and Surface Transportation Security Are Under Way, but Challenges Remain," October 16, 2007.

———, "Use of Covert Testing to Identify Security Vulnerabilities and Fraud, Waste, and Abuse," November 14, 2007.

———, "Airline Industry: Potential Mergers Driven by Financial and Competitive Pressures," July 2008.

———, "Federal Air Marshal Service Has Taken Actions to Fulfill Its Core Mission and Address Workforce Issues, but Additional Actions Are Needed to Improve Workforce Survey," January 2009.

———, "Efforts to Validate TSA's Passenger Screening Behavior Detection Program Under Way, but Opportunities Exist to Strengthen Validation and Address Operational Challenges," May 2010.

———, "TSA Has Enhanced Its Explosives Detection Requirements for Checked Baggage, but Additional Screening Actions Are Needed," July 2011.

U.S. House of Representatives/Committee on Transportation and Infrastructure, "Critical Lapses in Federal Aviation Administration Safety Oversight of Airlines: Abuses of Regulatory 'Partnership Programs,'" April 3, 2008.

U.S. House of Representatives/Investigative Report by the Committee on the Judiciary, "Plane Clothes: Lack of Anonymity at the

Federal Air Marshal Service Compromises Aviation and National Security," May 25, 2006.

U.S. House of Representatives/Majority Staff of the Committee on Homeland Security, "The State of Homeland Security 2007: Annual Report Card," 2007.

U.S. Intelligence Community/National Intelligence Council, "Mapping the Global Future," December 2004.

U.S. Office of Special Counsel, "Department of Transportation Report Substantiates Allegations of FAA Oversight Failures," July 22, 2010.

The White House, "Remarks by the President on America's Energy Security," March 30, 2011.

The White House Commission on Aviation Safety and Security, "Final Report of the Gore Commission on Aviation Safety and Security," February 12, 1997.

Supplemental

McGee, Bill, "On the Road/Behind the Screen" travel column, USA Today.com, 2003–2011 (numerous).

McGee, William J., *Air Transport World*, 1990–1997 (numerous).

———, *Condé Nast Traveler*, 2003–2009 (numerous).

———, "Case Closed?" *Condé Nast Traveler*, January 2005.

———, *Consumer Reports*, 2000–2011 (numerous).

———,"An Accident Waiting to Happen?" *Consumer Reports*, March 2007.

———, "Air Security: Why You're Not as Safe as You Think," *Consumer Reports*, February 2008.

———, *Consumer Reports Money Adviser*, 2004–2011 (numerous).

———, *Consumer Reports Online*, 2000–2011 (numerous).

———, *Consumer Reports ShopSmart*, 2006–2011 (numerous).

———(editor), *Consumer Reports Travel Letter*, 2000–2003 (numerous).

———, *Consumer Reports WebWatch*, 2002–2005 (numerous white papers).

———, "Sky Box: Shag or Shul?" *New York*, September 27, 1999.
———, "Forcing the FAA to Fly Blind," *New York Times*, April 10, 2011.
———, "Plane Talk," *T&E*, February 2006.
———, "All Aboard," *T&E*, October 2006.

Notes

NOTE: IN CERTAIN CASES FOLLOW-UP interviews were conducted in person and/or by telephone and/or by email.

In addition, there were dozens of other interview subjects who did not wish to be identified; these included FAA safety inspectors, air traffic controllers, federal air marshals, TSA screeners, airline mechanics, airline pilots, airline dispatchers, airline customer service agents, airline flight attendants, and travel/aviation journalists.

Selected Initial Interviews

Randy Babbitt [telephone], 1 Jul 2011.

Ritu Bararia, Gurgaon, India, 19 Apr 2011.

Arnold Barnett, Cambridge, Massachusetts, 1 Apr 2011.

Helane Becker [telephone], 1 Jun 2011.

Gordon Bethune [telephone], 8 Jun 2011.

Larry Bleidner, Los Angeles, California, 26 Mar 2011.

Jack Blomfeld [telephone], 22 Aug 2011.

Severin Borenstein, Berkeley, California, 28 Mar 2011.

David Bourne, Washington, D.C., 9 Jun 2011.

Susan Bourque, Buffalo, New York, 2 Jul 2011.

Joe Brancatelli [telephone], 23 May 2011.

Chris Brown [telephone], 3 Aug 2011.

Fred Brown [telephone], 16 Mar 2011.

Jan (Lohr) Brown [telephone], 11 Mar 2011.

Gabe Bruno, Arlington, Virginia, 8 Jun 2011.
Linda Burbank, San Francisco, California, 29 Mar 2011.
Brenda Cantrell, Scottsboro, Alabama, 22 Jan 2011.
Al Carlo [telephone], 19 Jul 2011.
Dave Carroll [telephone], 9 Feb 2011.
George Castro, Atlanta, Georgia, 20 Jan 2011.
Gordon Clark [telephone], 20 Jun 2011.
Jenny Cobb, London, U.K., 20 Apr 2011.
Roger Cohen [telephone], 9 Mar 2011.
Bill Collins, Tulsa, Oklahoma, 29 Jun 2011.
John Conley [telephone], 14 Jul 2011.
Jerry Costello [telephone], 21 Jul 2011.
Jami Counter, New York, New York, 25 Jan 2011.
Robert Crandall, Palm City, Florida, 7 Feb 2011.
Todd Curtis [telephone], 1 Jun 2011.
Mary Rose Diefenderfer [telephone], 16 Jun 2011.
Gail Dunham [telephone], 24 May 2011.
Jill Dunn [telephone], 12 Jun 2011.
Bogdan Dzakovic [telephone], 13 Aug 2011.
Karen Eckert, Buffalo, New York, 2 Jul 2011.
Gareth Edmondson-Jones [telephone], 2 Jun 2011.
Clark Kent Ervin [email], 9 Aug 2011.
Kim Farrington, Arlington, Virginia, 8 Jun 2011.
Linda Faulk, Denver, Colorado, 16 Jul 2011.
Jim Faulkner [telephone], 8 Jul 2011.
Charles Foss, St. Louis, Missouri, 29 Jun 2011.
Eadie Francis, San Diego, California, 22 Jul 2011.
Pat Friend [telephone], 18 May 2011.
Billy Glover, Renton, Washington, 19 Jul 2011.
John Goglia, Boston, Massachusetts, 1 Apr 2011.
Lisa Goldman, Jamaica, New York, 26 Mar 2011.
Bob Goldner, Washington, D.C., 6 Jun 2011.
Kenneth Goodpaster [telephone], 23 May 2011.

Mary Anne Greczyn [email], 8 Jun 2011.
Bill Hagan [telephone], 15 Apr 2011.
James Hall [telephone], 15 Jun 2011.
Kate Hanni [telephone], 14 Jun 2011.
Al Haynes [telephone], 18 Mar 2011.
John Heimlich [telephone], 3 Aug 2011.
Harris Herman, New York, New York, 25 Feb 2011.
Deborah Hersman, Washington, D.C., 21 Mar 2011.
Troy Hinrichs, Tulsa, Oklahoma, 30 Jun 2011.
Alex Hinton, Dallas, Texas, 30 Jun 2011.
George Hobica, New York, New York, 25 Jan 2011.
James Hoffa, Washington, D.C., 9 Jun 2011.
Hubert Horan [telephone], 4 Mar 2011.
Paul Hudson [telephone], 11 May 2011.
Ed Jeszka, Arlington, Virginia, 8 Jun 2011.
Jeff Johnstone [telephone], 2 Apr 2011.
Kori Blalock Keller, Washington, D.C., 7 Jun 2011.
Gary Kelly [telephone], 26 Aug 2011.
Louie Key [telephone], 2 Jun 2011.
John King, Tewksbury, Massachusetts, 23 Jun 2011.
Joan Kline, Morristown, New Jersey, 15 Jan 2011.
Kendall Krieg [telephone], 19 Jul 2011.
Susan Kurland, Washington, D.C., 6 Jun 2011.
Raymond LaHood, Washington, D.C., 6 Jun 2011.
Michael Landry, Tulsa, Oklahoma, 29 Jun 2011.
James Lardner [telephone], 11 Mar 2011.
Philomena Larsen, Buffalo, New York, 2 Jul 2011.
Robert MacLean [telephone], 29 Jul 2011.
Sarah MacLeod [telephone], 12 Aug 2011.
Pedro Mari, Tulsa, Oklahoma, 30 Jun 2011.
Nora Marshall, Washington, D.C., 21 Mar 2011.
Suzanne Martin, Atlanta, Georgia, 22 Jan 2011.
Bernadette McCulloch [telephone], 20 May 2011.

Mike McDonald, Tulsa, Oklahoma, 29 Jun 2011.

Dan McKenzie, Chicago, Illinois, 22 Jun 2011.

Diane McMullin, Renton, Washington, 19 Jul 2011.

Carol McVitty, Victorville, California, 22 Jul 2011.

Dawn Mikkelson [telephone], 27 May 2011.

P. L. Miller, San Diego, California, 22 Jul 2011.

Chris Moore [telephone], 25 Apr 2011.

Brian Moran [telephone], 10 Jun 2011.

Diana Moss [telephone], 17 Jun 2011.

Brad Mueller, Tulsa, Oklahoma, 29 Jun 2011.

Scott Mueller [telephone], 9 Jul 2011.

Rose Mulligan, Hartford, Connecticut, 12 Jun 2011.

Ralph Nader [telephone], 14 Jun 2011.

Lisa Orman [telephone], 15 Jun 2011.

Ms. Patel, Mumbai, India, 19 Apr 2011.

Barbara Peterson, Hastings, New York, 7 Mar 2011.

Bob Peterson, Renton, Washington, 19 Jul 2011.

Nicole Piasecki, Renton, Washington, 19 Jul 2011.

Peggy Post [telephone], 24 Feb 2011.

Joe Prater, Tulsa, Oklahoma, 29 Jun 2011.

Jon Prentice, Atlanta, Georgia, 22 Jan 2011.

Howard Putnam [telephone], 5 May 2011.

Joe Reynolds, Boston, Massachusetts, 25 Jun 2011.

David Saucedo [telephone], 22 Apr 2011.

Mitchell Schare, Hempstead, New York, 15 Feb 2011.

Bridget Ann Serchak, Washington, D.C., 21 Mar 2011.

Earlene Shaw [telephone], 26 May 2011.

Rolfe Shellenberger [telephone], 18 Feb 2011.

R. Bruce Silverthorn, Buffalo, New York, 1 Jul 2011.

Steven Slater [telephone], 27 Jun 2011.

Brett Snyder [telephone], 4 May 2011.

Jay Sorensen [telephone], 27 May 2011.

Tess Sosa [telephone], 4 Apr 2011.

Brent Strickland, Tulsa, Oklahoma, 30 Jun 2011.
Brian Sullivan [telephone], 12 Aug 2011.
William Swelbar [telephone], 28 Jun 2011.
Kerry Thorson, Tulsa, Oklahoma, 30 Jun 2011.
Glenn Tilton [telephone], 23 Jun 2011.
David Toomey [telephone], 10 Jun 2011.
Corky Townsend [telephone], 19 Jul 2011.
Maria Trejos [telephone], 26 May 2011.
Charles Unger [telephone], 15 Jun 2011.
Stephen Van Beek [telephone], 14 Mar 2011.
William Voss, Alexandria, Virginia, 6 Apr 2011.
Shainon Vyas, Mumbai, India, 18 Apr 2011.
Anthony Willett, Washington, D.C., 6 Jun 2011.
Jan Wood [email], 28 Apr 2011.
Richard Wyeroski, Arlington, Virginia, 8 Jun 2011.
Jerry Yates, Tulsa, Oklahoma, 29 Jun 2011.
Jeanne Yu, Renton, Washington, 19 Jul 2011.
Irene Zutell [telephone], 16 Jun 2011.

Chapter 2: What Happened to the Airlines?

1. Lardner and Kuttner, "Flying Blind: Airline Deregulation Reconsidered."
2. Author interview with Michael Levine.
3. Author interview with Robert Roach.
4. "Furious, American's Unions Talk of New Votes," *New York Times*, April 22, 2003.
5. American Customer Satisfaction Index, "Passenger Discontent Pervasive for Airlines."

Chapter 3: Collusion and Confusion: How Airlines Don't Play by the Rules—and How Passengers Pay

1. OpenSecrets.org, 2011.

2. Business Travel Coalition, "Predatory Airline Behavior Is Back."
3. Transportation Research Board, "Entry and Competition in the U.S. Airline Industry."
4. Ibid.
5. Oster and Strong, "Predatory Practices in the U.S. Airline Industry."

Chapter 4: So You Think You've Found the Lowest Fare

1. Bailey, "Air Transportation Deregulation."
2. Author interview with Bob Harrell; author interview with John Heimlich.
3. Author interview with Ed Perkins.
4. Author interview with Al Anolik.
5. Moss, "Airline Mergers at a Crossroads."
6. Author interview with Randy Petersen.
7. Author interview with Tim Winship.

Chapter 5: A Mad Race to the Bottom: How Airlines Mistreat Employees, Outsource, and Ignore Passengers

1. U.S. Intelligence Community/National Intelligence Council, "Mapping the Global Future."
2. Author interview with Pat Friend.
3. Ibid.

Chapter 6: When Your Airline Isn't Your Airline: Regional Carriers Provide Lower Levels of Service and Safety

1. Author interview with Bob Mann.
2. "Airline Ticketers Find Ways to Skirt New Law," *Buffalo News*, Dec. 13, 2010.
3. Author interview with Tim Winship.
4. Author interviews with Kevin Mitchell and Charlie Leocha.
5. Author interview with Juan Alonso.

6. Author interview with Todd Curtis.
7. Author interview with Michael Levine.

Chapter 7: Outrageous Outsourcing: The Single Greatest Threat to Airline Safety

1. U.S. Department of Transportation/Office of Inspector General, "Air Carriers' Outsourcing of Aircraft Maintenance."

Chapter 8: Unsafe at Any Altitude? Facing Unprecedented Dangers

1. Author interview with Todd Curtis.
2. Author interview with Jeff Marcus.
3. Author interview with Tom Haueter.
4. Author interview with Congressman James Oberstar.
5. Author interview with Tom Brantley.

Chapter 10: Lights, Camera, Strip Search: The Tragicomedy of Airline Security

1. Levin, "Airlines Seek to Move Air Marshals from First Class."
2. Author interview with Bruce Schneier.
3. U.S. Government Accountability Office, "Use of Covert Testing to Identify Security Vulnerabilities."
4. Peterson, "Inside Job: My Life as an Airport Screener."
5. "GAO Cites TSA Detection Equipment Standards Disarray," Representative John Mica, July 12, 2011.
6. *Canadian Journal of Police and Security Services*, June 2005.
7. "Technology and Liberty: Airport Security," American Civil Liberties Union, Nov. 17, 2010.

Chapter 11: Cloudy Skies: Aviation's Carbon Footprint

1. Author interview with Paul Dempsey.
2. Author interview with Juan Alonso.
3. Author interview with Ian Waitz.

4. Treehugger.com, "Greenopia Rankings Push Airlines to Pursue Virgin," May 20, 2011.
5. Author interview with Dudley Curtis.
6. Alonzo, Waitz, et al., "Greening U.S. Aviation."

Chapter 12: Ink-Stained Wreckage: How Airlines Manipulate the Media

1. Crouch, *The Bishop's Boys: A Life of Wilbur and Orville Wright.*
2. Author interview with Kevin Mitchell.
3. "Rules and Restrictions Apply," *Washington Post*, May 29, 2000.

Index